Chemical Structure
Software for
Personal Computers

Chemical Structure Software for Personal Computers

DANIEL E. MEYER, EDITOR
Advanced Research Technologies, Inc.

WENDY A. WARR, EDITOR
ICI Pharmaceuticals

———

RICHARD A. LOVE, ASSOCIATE EDITOR
American Chemical Society

ACS Professional Reference Book

AMERICAN CHEMICAL SOCIETY   WASHINGTON, DC   1988

Library of Congress Cataloging-in-Publication Data

Chemical structure software for personal computers/
editors, Daniel E. Meyer and Wendy A. Warr; associate
editor, Richard A. Love.

(ACS professional reference book)

p.   cm.
Bibliography: p.

Includes index.
ISBN 0-8412-1538-3. ISBN 0-8412-1539-1 (pbk.)

1. Chemical structure—Software—Catalogs.
2. Chemical structure—Software—Directories.
3. Microcomputers—Programming.

I. Meyer, Daniel E., 1956-   . II. Warr, Wendy A.,
1945-   . III. Love, Richard A.

QD471.C52   1988
541.2′2′02855369—dc19   88-8153
                                          CIP

About the Editors

Daniel E. Meyer is founder and president of Advanced Research Technologies, Inc., which provides consulting services to organizations in the information and scientific communities. He received a B.S. in zoology from the University of Maryland (College Park).

Meyer has held various professional positions including Associate Editor for *Archives of Environmental Health* (Heldref Publications), online search specialist for chemical and biomedical databases (Tracor Jitco), and Manager of New Product Development at the Institute for Scientific Information, primarily for the Chemical Information Division. He has been at the forefront of promoting the use of personal computer software for handling chemical-structure data and has presented more than 20 papers at international conferences on this topic. He has published more than 10 articles and book chapters on chemical information systems and PC applications.

Meyer serves on the Program Committee of the ACS Division of Chemical Information and the Program Committee of NFAIS (an organization of database producers) and has recently been selected to serve as a member of the ACS Software Advisory Board.

Wendy A. Warr has M.A. and D.Phil. degrees in chemistry from the University of Oxford, England. She first became involved with information systems in 1968 through the Experimental Information Unit at Oxford, where she worked as a part-time research assistant while doing her doctoral research in Organic Chemistry. She was one of the first people in England to use Wiswesser line notation for encoding chemical structures.

After a brief spell in chemical research and development in industry, she joined ICI Pharmaceuticals in 1972 and has worked there ever since, progressing from information scientist to Senior Systems Analyst and Project Manager and then back into an enlarged Information Services Section, which she now manages.

Warr is a member of numerous professional organizations including the American Chemical Society and the Royal Society of Chemistry. She is on the Program Committee of the ACS Division of Chemical Information and is a member of the editorial board of the Journal of Chemical Information and Computer Sciences and the advisory board of ACS Books.

Her particular interests in chemical information are substructure searching, graphics, end-user searching, applications of microcomputers, online chemical databases, and front-end systems. She is editor of *Graphics for Chemical Information: Integration with Text and Data*, ACS Symposium Series No. 341, published in 1987, and *Chemical Structures: The International Language of Chemistry*, published by Springer Verlag in 1988.

Richard A. Love did his undergraduate work in chemistry at the University of California, Los Angeles, and holds a Ph.D. in Inorganic Chemistry from the University of Southern California. Following his graduate studies and a return to UCLA for postdoctoral work, he worked for several years as research scientist in organometallic chemistry at W.R. Grace & Company. As Senior Research Associate at the Publications Division of the American Chemical Society, he has responsibilities in the design and implementation of online full-text databases,

machine-readable manuscripts from authors, scientific application software for personal computers, and electronic publishing.

Contributors

Daniel E. Meyer
President
Advanced Research
  Technologies, Inc.
P.O. Box 556
Wayne, PA 19087

Wendy A. Warr
Manager, Information Services Section
ICI Pharmaceuticals
Mereside, Alderley Park
Macclesfield, Cheshire SK10 4TG
England

Richard A. Love
Senior Research Associate
Publications Division
American Chemical Society
1155 Sixteenth Street, N.W.
Washington, DC 20036

Cyrelle K. Gerson
Manager, Media Courses
Continuing Education Department
American Chemical Society
1155 Sixteenth Street, N.W.
Washington, DC 20036

William G. Town
Managing Director
Hampden Data Services Ltd.
167 Oxford Road
Cowley, Oxford OX4 2ES
England

Contents

ix

Preface

The primary intent of this book is to serve as a directory of currently available software products for creating and using chemical-structure diagrams on a personal computer. The book concentrates almost exclusively on software for the IBM and Macintosh computers.

We have an ongoing interest in PC-based chemical-structure software and have published journal articles and presented papers at national conferences on the topic. Although we routinely use and have reviewed many of these programs, we have compiled the product descriptions from information supplied by the software producers on author-designed forms. The producers were asked to describe their software product, to specify the required hardware (personal computer, minimum RAM, and supported peripherals), and to give the price of their product(s). For the chapters on structure-drawing and molecular-modeling software (Chapters 2 and 5), the producers were requested to use their own program(s) to print, for publication, a test chemical structure. For the individual product descriptions, we have grouped hardware, such as personal computers and printers, into broad categories according to functional equivalence. For example, the phrase "IBM PCs and compatible systems" is used to include all PCs that operate under PC-DOS and MS-DOS, which would include all PC/XT/AT-class machines.

We are interested in maintaining up-to-date files on chemical-structure software, and we will appreciate notices and brochures on new and updated programs in this area from interested readers and software producers. Because many of the software products listed in this book are actively updated, we have described the most current version of the software at the time of submission for publication (June 1988). In future editions of this book, we will incorporate the new features that distinguish the most current release of the software discussed in this volume, and we will include new products that have been introduced since this edition.

Although we have substantial experience in evaluating software for personal computers, we have not necessarily tested every product described here, and we accept no responsibility for missing functionalities or erroneous claims. We tried to include all relevant software and regret the omission of any programs that we were not aware of at the time of publication.

We thank the many people who made this book possible, especially Maureen Rouhi, Janet Dodd, and Robin Giroux of the ACS Books Department.

DANIEL E. MEYER
WENDY A. WARR
RICHARD A. LOVE

August 1988

CHAPTER ONE

Current Status of Computer-Assisted Drawing of Chemical Structures

Wendy A. Warr and William G. Town

Progress in application software has moved in tune with developments in hardware and operating environments. The first generation of computers, produced and used from 1938 to 1953, was based on relays and valves and had virtually no operating software. The second generation (1953–1963) was based on transistors and batch-processing operating systems. Third-generation computers (approximately 1963–1973) were equipped with small- and medium-scale integration, an innovation that made possible the development of interactive computing systems. Fourth-generation computers (1973–1983) have large-scale and very large scale integration. The resulting increase in computing power led to more-complex operating environments involving networking and workstations. We are now in the era of fifth-generation computers (1983–1993?) with very large scale integration and parallel processors, and the current emphasis is on the development of intelligent knowledge-based systems.

The effect of this evolution of hardware and operating systems may be observed in chemical-structure information systems. A typical software application in this area in a batch-processing environment was CROSSBOW, which appeared in 1967 (1). Software of the interactive era, which required teletype-compatible terminals, is typified by the development of the National Institutes of Health/Environmental Protection Agency Chemical Information System (CIS) (2). This kind of software, through its use of character and matrix graphics on teletype terminals, made chemical information systems available to laboratory scientists. In the early 1980s, interactive graphics systems such as DARC (*3*) and MACCS (*4*) and the publicly available CAS ONLINE (*5*) and DARC/Questel services began to appear.

At present, in the distributed processing environment of the later 1980s, new types of systems are being made possible by the advent of personal computer (PC)-based software. One application of this kind of software is the integration of text, chemical structures, mathematical equations, graphs, charts, and spectra into a single document. Software is also becoming available for the maintenance of chemical structure databases on PCs and for offline query formulation for online graphics systems.

Ultimately, the PC will provide the graphics interface not only to PC databases but also to company databases on a mainframe and to public databases on online hosts. With this software on a personal workstation, the chemist will be able to create queries and send them to PC-resident, company, or public files. In turn, the chemist will be able to download chemical structure information into personal databases for storage and analysis or for publication.

Workstations and Personal Computers

With the remarkable processing power of the third-generation microprocessors and the improved graphics-processing capabilities found in PCs today, there is an ever-increasing convergence between PCs and scientific workstations.

First-generation microcomputers were relatively slow and had low-resolution graphics screens. Many microcomputers had no graphics capabilities at all. In general, they were used by individuals for local processing and were not linked as terminals to mainframes or local networks. With the evolution of terminal emulation software and communication boards that facilitated these links, microcomputers replaced "dumb" terminals as the preferred method of connecting to mainframe computers. Soon after, networking software and hardware made possible the grouping of microcomputers into local area networks. However, for computationally intensive tasks requiring high-resolution graphics, such as molecular modeling and computer-aided design, the PC was inadequate, unless it was enhanced by special graphics cards, screens, and additional coprocessors. The scientific workstation fills this gap.

The scientific-workstation market is relatively young and has evolved in parallel with the PC market. Scientific workstations are characterized by high resolution screen raster (e.g. Silicon Graphics IRIS workstations) or vector (Apollo Domain) graphics. These workstations often offer a choice of up to 256 colors, window-oriented user environments, and multitasking operating systems, and many of these are UNIX-based systems. Also, these workstations often have high-bandwidth network links (such as Ethernet) to other workstations or mainframes and shared computational facilities (e.g., array processors or the Weitek tiling engine).

Third-generation microcomputers, which are based on 32-bit micro-processors, have the computational power of a minicomputer and larger memory address capacities (up to 4 gigabytes) than their predecessors. Examples of these new systems are the IBM PS/2 and Apple Macintosh II computers equipped with the Intel 80386 or Motorola 68020–68030 chips. These computers can process about 4 MIPS (million instructions per second), which is comparable to the processing power of the minicomputer DEC VAX 11/780. Microcomputers offer high-resolution screen images through proprietary processors or through one of the new graphics coprocessor chips, for example the Intel 82786 or the Texas Instruments TMS 34010. The price of these PCs is about an order of magnitude lower than that of the typical scientific workstation several years ago.

Scientific-workstation manufacturers have responded with lower prices and newer models that directly compete with the top-range microcom-puters. Sun, for example, recently announced a new low-cost model that supports both PC-DOS and UNIX. Mainframe and minicomputer compa-nies, such as DEC, have also entered the scientific-workstation market with offerings such as the VaxStation 2000. These workstations are compatible, at the operating-system level, with the entire VAX range of computers. As new operating systems, such as the OS/2 with its Presentation Manager, reach their full potential and as the IBM move to Standard Applications Architecture for its microcomputer-to-mainframe computers becomes a reality, the emphasis on networking and integration will increase. It is certain that the software systems that are best able to adapt to this challenge will be the systems of choice for the future.

Compact Disk Read Only Memory

Storage capacity at the PC level has steadily increased. The new compact disk read only memory (CD-ROM) and write once read many (WORM) peripheral disk drives offer very large storage capacities suitable for archiving information. WORM disks can store up to 200 megabytes of data, compared with 500 megabytes for CD-ROM. The enormous storage capacity of CD-ROM is particularly useful for storing graphical data which have high storage requirements.

As a medium for storing bibliographic databases within a local environment, CD-ROM is best suited for information that is accessed on a regular basis and that does not age too rapidly. The time delays and costs involved in mastering and replicating new versions of a database have the consequence that the information can not always be current. However, the local user of a CD-ROM can search as often and take as long as required because there are no additional costs once the CD-ROM has been purchased. Downloading and postprocessing of search results are also possible without a large financial penalty.

Chemical-Structure Software

Five categories of software for handling chemical structures electronically have been defined.
• Structure-drawing software
• Graphics terminal emulation software
• Structure management software
• Software for three-dimensional molecular graphics and modeling
• Special-application software

Structure-Drawing Software

Software for drawing and displaying chemical structures on PCs is a growth area. Graphics for chemical structures and integration of structures with text and data is the subject of a recent book (6). An overview of this class of software for technical manuscript preparation and a comparison of technical word processors for scientific writers has been published recently (7).

A fundamental distinction can be drawn between software packages that are "image only" and those that are multipurpose. Many of the graphics packages currently on the market have been developed solely for document preparation. These packages are not based on connection tables and therefore do not have substructure search capabilities. Many of these programs use either character-based or graphics interface techniques to produce structure diagrams. These packages can, in turn, be divided into two classes. WYSIWYG (what you see is what you get) programs require the computer to operate, at least part of the time, in graphics mode. Mark-up language programs allow the computer to operate in the text mode. Some of the software packages can also produce mathematical symbols and expressions. Indeed, some of these programs were specifically designed for scientific word processing, and they handle chemical structures only with difficulty.

Graphics Terminal Emulation Software

The first area in which PCs made an impact on chemical information retrieval was that of graphics communication software, also known as graphics terminal emulation software. Most of the common graphics terminals (e.g., Tektronix 4010 and VT 640) can be emulated on more than one type of PC. This class of software is fundamentally different from most of the chemical-structure-handling software detailed in this book. These products have probably revolutionized the way in which we access chemical information. Until emulation packages of this sort became available, the only way of using services like CAS ONLINE or DARC/Questel in graphics mode was to use very expensive graphics terminals. The terminal emulation packages offer a cheap alternative and, as a result, have increased access to these and other online services that provide graphics information.

Structure Management Software

The structure management software packages discussed in this book differ from each other in many respects, but they all offer a substructure search capability. This ability to search for structures and structural fragments is made possible by the storage of the structures in connection table format rather than as vector or bit-mapped files. Most of the packages discussed also include the ability to verify the chemical accuracy of an input structure. Some also provide start-up chemical-structure databases and can do data calculations or database management.

Software for Three-Dimensional Molecular Graphics and Modeling

Several PC programs in the market are specifically designed for drawing chemical structures followed by production of a three-dimensional (3-D) diagram for input to molecular modeling. The molecular modeling itself may or may not take place within the context of the program or even in the PC. The structure input packages differ from others described in this book because there is no need for high-resolution graphics at the input stage or for production of publication-quality documents. However, there are obvious advantages in the use of 3-D templates at the structure input stage, and some of the packages offer initial template files.

Special-Application Software

A few software packages allow graphics input of chemical structures and are designed for very narrow applications. They may assist users in accessing reference sources that use specialized codes or classification conventions. These products are discussed in Chapter 6.

Offline Formulation of Structure Query

Uploading of chemical structures is a special application. To understand the advantages that uploading gives users, consider how online graphics systems are currently used. When a user (chemist or information specialist) constructs a query for a structure search in the CAS ONLINE or DARC/Questel systems, there is interaction with a remote mainframe computer. The user sends commands through a telecommunication network, and in turn, the remote computer interprets those commands and constructs the structure image for display at the user's terminal.

Advantages can be gained if the structure diagram were to be created locally at the user's PC. This process would reduce the load on the host computer, reduce telecommunication traffic, and reduce costs for the user. Also, the stress on the user caused by the need to remember all the necessary commands for creating the diagram while the costs are ticking up (the so-called "taxi meter" syndrome) would be reduced. By formulating searches off line, the user has time to think about the query and be assured that the query is correct before the user logs in and uploads the query to the host computer. The vendor of the database accomplishes this interface by defining standard formats for queries that match the requirements of the user's software and the host system.

Software for offline formulation of a structure query and uploading is often referred to as a "front end". An early example of front-end software is SuperStructure (8) that was produced by Fein Marquart Associates of Baltimore, Maryland, to access the Chemical Information System (CIS) (2). New front-end software is becoming available that accesses the Beilstein, DARC/Questel, and STN International databases.

Downloading of Chemical-Structure Information

The converse of offline query formulation and uploading a search is downloading information. At present, many packages are available that assist users in interacting with text-based systems. Some of these packages allow the user to capture text from a public system. Corresponding developments in the area of chemical-structure information will occur. It will then be possible to download structure information that has been retrieved from a public file into the PC so that the information can be further processed. Once online hosts make connection tables available to PC-resident software, local search of the downloaded chemical-structure information will also become available.

One downloading facility that is already in use is the capture of bit-mapped and vectorized graphics from online hosts. These graphics can be stored, redisplayed, and printed, but in most cases, there is no way of taking the captured image of a structure and converting the image into a connection table.

Conclusion

The evolution of chemical information systems has been strongly influenced by the available computer hardware. Chemists communicate structure-related concepts by means of conventional two-dimensional diagrams, and early computer systems were ill adapted to meet these descriptive needs. The advent of graphic-structure input and output on PCs has opened up new possibilities for chemists. Unfortunately, the chemist is also faced with many competing products for doing chemical-structure drawings. Although there are some advantages in having a large number of software solutions available, there are obvious disadvantages if the user has to purchase and learn many different packages—one for producing reports and publications, another for accessing online systems, a third for substructure search of company databases, and so on. The chemist may also find that a preferred method of structure drawing will not integrate with the company's choice of word-processing or office automation software. The key area for future development in this industry must inevitably be interfacing and integration.

References

1. Eakin, D. R. In *Chemical Information Systems;* Ash, J. E.; Hyde, E., Eds; Ellis Horwood: Chichester, United Kingdom, 1974; pp 227–242.
2. Feldmann, R. J. In *Computer Representation and Manipulation of Chemical Information;* Wipke, W. T.; Heller, S. R.; Feldmann, R. J.; Hyde, E. Eds; John Wiley: New York, 1974; pp 55–81.
3. Attias, R. *J. Chem. Inf. Comput. Sci.* **1983,** *23(3),* 102–108.
4. *Communication, Storage and Retrieval of Chemical Information;* Ash, J. E.; Chubb, P.; Ward, S. E.; Welford, S.; Willett, P., Eds; Ellis Horwood: Chichester, United Kingdom, 1985; pp 182–202.
5. Farmer, N. A.; O'Hara, M. P. *Database* **1980,** *3(4),* 10–25.
6. *Graphics for Chemical Structures: Integration with Text and Data* ; Warr, W. A. Ed.; ACS Symposium Series 341; American Chemical Society: Washington, DC, 1987.
7. Gerson, C. K.; Love, R. A. *Anal. Chem.* **1987,** *59(17),* 1031A–1048A.
8. McDaniel, J. R.; Fein, A. E. In *Graphics for Chemical Structures: Integration with Text and Data;* Warr, W. A. Ed.; ACS Symposium Series 341; American Chemical Society: Washington, DC, 1987; pp 62–79.

CHAPTER TWO

Structure-Drawing Software

Richard A. Love

C hemical-structure diagrams are essential tools of the chemical profession. In chemistry publications, lectures, and grant proposals, one can usually find a reaction scheme or diagram that succinctly illustrates the ongoing discussion of chemistry. A well-drawn structure often tells more about the behavior of a molecule than several paragraphs of descriptive text can. Depending on the care with which a chemical structure is drawn, a chemical-structure drawing can succinctly depict how a molecule bonds internally, how it interacts with other molecules, or how it behaves in three-dimensional space.

Computer-Based Drawing of Chemical Structures

Many shortcuts are used when chemical structures are drawn. For example, it is common to leave out specific notations for carbon and hydrogen atoms by drawing an organic compound with stick bonds and to use a circle within a regular polygon to represent aromaticity. These representations have become accepted icons by chemists who use them in preparing reports and manuscripts or for simple musings on the chalk board.

The preparation of chemical-structure diagrams by hand for publication, whether by a chemist or a graphics artist, is a labor-intensive process

1538–3/88/0009/$08.00/0 © 1988 American Chemical Society

involving templates, sharp-pointed pens, and strategic spots of "white out" before the artwork is cut and pasted into the final draft for publication.

With the proliferation of microcomputers in academia and industry, numerous software programs have been developed that create chemical-structure diagrams. The advantage of using software over the traditional method of drawing structures by hand is that software stores the diagrams in an electronic form, can alter them in any way (within the constraints of the program), and reprints them as new originals whenever required. Many programs directly integrate the drawings into the text of a document. This integration allows the user to make corrections and changes to the text without having to worry about redoing the graphic illustrations. Chemists are demanding greater sophistication in drawing packages as they realize that the quality of their drawings greatly enhances the understanding of their presentations.

Having the diagrams in electronic form also allows multiple use of the information. For example, a structure stored as a connection table possesses a logical format that allows other programs, such as a database manager (*see* Chapter 4), to form the structures into a searchable database for later recall. The structure could also be used as an input to other programs, such as a molecular modeler (*see* Chapter 5). The electronic form of the diagram gives greater dimension and portability to the information; once a structure is created, it can be transferred between different applications and computer systems for other uses.

Computer Systems for Drawing Chemical Structures

Most chemical-structure-diagram programs for microcomputers on the market are supported by either the IBM PCs and compatible systems or the Apple Macintosh systems. The two exceptions are CHEM, which was developed for the UNIX operating system, and GIOS, which was developed for the Apple II and Apple IIc systems. The following discussion compares these two environments.

Apple Macintosh System

Two Macintosh drawing programs that are discussed in this chapter are ChemDraw and ChemIntosh. In the Macintosh environment, the user sees a consistent program interface because of the standards recommended by Apple Computer to the software developers. This consistency results from the Macintosh system having manager programs residing within the computer ROM that software developers use to connect their application programs with the operating system.

One of the managers used by graphics programs, such as structure-drawing packages, is the QuickDraw Manager. Simple shapes, such as regular polygons (triangles, squares, pentagons, etc.), circles, or lines, are defined by the software developers as simple QuickDraw primitives. For example, with the QuickDraw notation, the instruction "Oval(0,0,5,10)" defines a primitive that connects the Cartesian coordinates (0,0) and (5,10) with an oval. Freely drawn forms are specified by simple-line primitives. Macintosh software developers have an incentive to make use of the QuickDraw Manager because it assures them that their graphic images can be ported into other Macintosh programs, such as a word processor, if the program also conforms to the standards. Virtually all do. Also, the graphics interface encouraged by QuickDraw greatly enhances the use of the mouse attachment that is standard with the Macintosh computer. The drawing primitives are readily incorporated in the user interface as icons that can easily be accessed and moved around by the user at a few clicks of the mouse.

Macintosh application programs draw images on the screen or print them on the printer by sending the QuickDraw instructions to the appropriate manager. The QuickDraw Manager converts these instructions into the correct pixels for screen display. Having the program in ROM gives impressively fast screen display on the Macintosh. To display the image on the output device, the print manager sends the QuickDraw instructions to the appropriate driver for the device, whether it be a laser printer, a dot matrix printer, or a plotter. For instance, a PostScript driver, written by the printer hardware manufacturer, interprets the QuickDraw instructions for printing the image on PostScript printers.

Another standard that Apple Computer has successfully recommended to software developers is the PICT storage format for graphic images. All graphic images that are passed from one application package to another via the Macintosh "clipboard" feature are passed by storing the transferred images in the standard PICT format. The application packages accomplish this transfer by filtering their QuickDraw instructions through the QuickDraw Manager, which then writes out the PICT file to the clipboard. Two products for the Macintosh, ChemPlate and DrawStructures, are libraries of chemical structures that are stored in the PICT format and can be used with object-oriented drawing programs, such as MacDraw and SuperPaint, to build chemical-structure diagrams.

The strengths of the Apple Macintosh environment derive from the standards and ROM-resident managers. The user sees a consistent interface for different applications, a feature that makes the learning process much easier. Once a user becomes accustomed to a few programs, the use of new ones is almost intuitive. The user can also work with different application programs more easily. For example, the font manager allows the user to use fonts purchased for one application for other applications on the Macintosh. Thus, Organic Fonts and the Hopkins font can be used with any Macintosh word processor to generate organic chemical structures by typing out the discrete components of the structure.

Software developers also benefit from the recommended standards. Many resources are available to them even before their applications are written. For example, developers need not know anything about PostScript to print with the Apple LaserWriter (a PostScript printing device); they only need to know about QuickDraw, and the Macintosh managing environment takes care of the rest. This standardization may be one reason that prices are consistently lower for Macintosh products compared with comparable products with the DOS operating system.

IBM PC System

The IBM PC environment is not the same as the Macintosh environment from the points of view of the users and developers. IBM architecture is more open and does not have consistent standard interfaces for the developers. Consequently, developers often start from scratch in building programs, and different products have less chances of communicating with each other. From the perspective of the user, this lack of communication gives a fractured environment and limited opportunities for porting information between applications. Software developers usually solve this problem by writing specific interfaces to bridge different applications. For example, developers of chemical-structure-drawing packages need to write an interface for each specific word processor they choose to support in order to integrate their drawings directly into a working document created by the word processor. As a result, the most popular word-processing packages tend to be supported, and overall development costs for individual packages increase.

In general, software developers for the IBM environment take two approaches for the inclusion of molecular-structure diagrams in a document.

1. The structure-drawing capabilities are developed with a word processor as a single integrated package.

    Programs such as ChiWriter, Spellbinder Scientific, T³, TechSet, and The Egg accomplish this integration by providing a full-function scientific word-processing package that is capable of drawing chemical-structure diagrams. Many of these programs may also have other features, such as mathematical-text handling or foreign alphabets, that extend the capabilities of the scientific word processor.

2. The structure-drawing capabilities are developed as a separate package.

The structures are called into supported word-processing applications at the time of screen display or printing. For example, ChemText can show structures integrated with the text directly on the screen. ChemText accomplishes this integration by dynamically calling in and displaying the structures, which are stored in a separate file. To create and edit the structure, the user accesses the structure file by using the Molecule Editor program provided with the software. Other packages, such as Molecular Presentation Graphics (MPG), insert the structure directly into a working document by having the user place a unique call-out code in the word-processing file and use an external formatter to print the document.

In addition to these two approaches for displaying and printing chemical-structure diagrams, developers of software for IBM PCs and compatible computers are using several methods to create diagrams.

1. Graphics interface

This approach closely imitates the Macintosh environment that uses a mouse and a true graphics interface to construct chemical-structure diagrams. Examples of programs in this category are ChemText, MPG, PsiGen, and WIMP. These packages are often described as being WYSIWYG (What You See Is What You Get), because the graphic that is constructed on the screen corresponds very closely with the final printed image.

2. Character- or font-based method

This is the approach developed for the majority of packages that have full-function word-processing features. These programs treat the atoms and various bond types as single characters or fonts in much the same manner that the regular text and other special characters are produced from fonts. To use these programs, a user builds a structure one character at a time on the screen until the desired structure is complete. This somewhat tedious task is made easier in many of these packages by a macro feature that can automatically repeat the required key sequences for a particular structure. Examples in this category are ChiWriter, Spellbinder Scientific, VolksWriter Scientific, T^3, TechWriter, The Egg, and WordMARC Composer+. The character- or font-based packages are also often described as being WYSIWYG.

3. Mark up

   This is the approach for two packages discussed in this chapter, TechSet and CHEM. (CHEM is not an IBM PC-compatible or DOS program. It is only available under the UNIX operating system.) The mark-up approach uses in-line codes that describe the drawing as a set of mnemonic instructions. The advantage of this approach is that the codes are written in the standard ASCII character set so that the user is free to use any word processor and computer operating system that can write out a clean ASCII file; most can. The mark-up files are displayed on or printed by the appropriate screen or printer drivers.

The fractured environment for the IBM may become more unified as more developers take advantage of operating systems such as Microsoft Windows. Microsoft Windows emulates the Macintosh operating system and gives developers many standard tools with which to create a consistent user interface for their products. Microsoft Windows has effectively been supported by the new IBM OS/2 Presentation Manager, which is also a windowing environment and allows developers to easily port their Window applications into Presentation.

When choosing a chemical-structure-drawing program, one issue to consider is the quality of the printed image. One of the primary reasons that the chemical community is requesting this kind of software is for it to acquire the ability to create chemical-structure diagrams for publication. The higher the quality of the printed image, the better it can be reproduced for publication.

In general, three types of output printing devices are supported by these packages: dot matrix, laser, and high-end laser imagers. By far, the laser imagers give the highest quality output, in the order of 1000–2500-dpi (dots per inch) resolution. But these printers are expensive, and few users have access to them. Indeed, these are the same printers used to print documents for journal or book publication. Laser printers and a few dot matrix printers can also give high-quality images in the range of 300–600 dpi.

For IBM packages, this quality is a function of the ability of the software developer to write the appropriate driver so that the printer can take advantage of the maximum resolution of the printer. All Macintosh packages and a few IBM packages, notably ChemText and WIMP, give consistently excellent images because they access PostScript printing devices, such as the Apple LaserWriter (300-dpi resolution), Varityper 600 (600-dpi resolution), and Linatronic 300 (a high-end laser imager with 2540-dpi resolution). Many IBM and Macintosh products also support plotters that give a very high quality, in effect, infinite, resolution for the printed image.

Product Selection

The packages described in this chapter were chosen on the basis of their ability to create the test molecular structure given for *trans*-5-(2-hydroxy-phenyl)-1-(3-mercapto-1-oxopropyl)-D-proline.

Test Structure

An example of the ability of the program to create the structure is given with each product description. Programs that did not reproduce the test structure at a quality consistent with good presentation were not included. Several producers declined to submit a creation of the test structure.

Chemical-Structure-Drawing Programs

The programs for chemical-structure drawing that are discussed in this chapter are the following.
- CHEM (AT&T Bell Laboratories)
- ChemDraw (Cambridge Scientific Computing)
- ChemIntosh (SoftShell Company)
- ChemPlate/Hopkins (The Johns Hopkins University)
- ChemText (Molecular Design Limited)
- ChemWord (Laboratory Software)
- ChiWriter (Software Design)
- 2D-CHEMICAL STRUCTURES (VCH Wissenschaftliche Software)
- DrawStructures and Organic Fonts (Modern Graphics)
- GIOS (Georg Thieme Verlag)
- Molecular Presentation Graphics (Hawk Scientific)
- PICASSO (Fraser Williams)
- PsiGen (Hampden Data Services)
- Spellbinder Scientific (Spellbinder Software Products)
- T^3 (TCI Software Research)

- TechSet (Software Development & Distribution Center)
- TechWriter (CMI Software)
- The Egg (Peregrine Falcon Company)
- WIMP (Aldrich Chemical Company)
- WordMARC Composer+ (MARC Software International)
  The following related programs are discussed in later chapters.
- ChemTalk (*see* Chapter 3)
- ChemBase, PC-SABRE, and PsiBase (*see* Chapter 4)
- Chem-3D and ChemDraft (*see* Chapter 5)

Product Descriptions

 ## CHEM

CHEM is a program that runs under the AT&T UNIX operating system and is used to produce chemical-structure diagrams. The program is used with *troff*, the text-formatting program of AT&T for typesetting, and is similar to other *troff* preprocessors such as EQN (used to create mathematical equations) and TBL (used to create tables). Specifically, CHEM is a preprocessor for PIC, which is used to create schemes and diagrams. Like PIC, CHEM uses a natural language mark-up approach that attempts to capture the way a chemist might describe a molecule during a conversation. For example, methyl acetate would be described in CHEM by the following sequence: CH3, bond right, C, double bond, O, bond 120 from C, O, bond right, CH3. The CHEM program converts this mnemonic description into a sequence of PIC graphic language commands, which in turn are interpreted by the *troff* formatter. Numbers are automatically subscripted.

Like PIC and many other utility programs in the UNIX environment, CHEM gives the user the bare minimum of help and diagnostics. But if the user is comfortable with *troff* and its preprocessors, then CHEM would not present any greater challenge.

CHEM Output

CHEM is best at creating organic, bio-organic, and polymeric molecules. To draw molecular constructs not supported by the CHEM program, the user can use PIC directly. Because CHEM is part of *troff*, it is independent of input device and can be used to produce true publication-quality output on laser printers and on high-quality typesetting machines that support *troff*.

Hardware

- ☐ PC(s): any PC that supports AT&T UNIX System 4.2 or higher. The computer must have the *troff* and AWK programs. (CHEM is written in AWK.)
- ☐ Graphic board(s): not applicable
- ☐ Mouse(s): not applicable
- ☐ Printer(s): any printer or output device that supports the *troff* formatter.
- ☐ Minimum RAM: not applicable
- ☐ Dual-floppy-disk system: no

Price and Vendor

- ☐ CHEM is a public-domain program. For a source code listing of CHEM, see reference *1*.

 ## ChemDraw

ChemDraw is an Apple Macintosh application program that uses pull-down menus, icons, and a mouse. The program can create various bond types such as normal, wedged, hashed, dashed, squiggled, or multiple double and triple bonds. ChemDraw has stored templates for saturated and unsaturated rings of three to eight atoms. ChemDraw can optionally draw ring or bond fragments by using fixed bond lengths and can customize bond widths and lengths by using a preference menu. Special icons for creating arrows (including retrograde synthetic arrows) and arcs allow the user to draw reaction schemes. The program can draw circles

ChemDraw Output

and brackets and includes text tools for labels and captions and a number of other general-purpose drawing tools for aligning and framing drawn structures for display. Other features include the ability to rotate and flip structures containing atoms labeled about the *x* and *y* axes; to automatically draw zig-zagging acyclic chains; to show the entire page layout and use all of the drawing tools available with the document window; to write out structure in a connection table format; and to port drawn structures into a companion program, Chem3D (*see* Chapter 5). The output from ChemDraw is easily integrated into any Macintosh word-processing package, such as Microsoft Word, via the clipboard.

Hardware
☐ PC(s): Macintosh
☐ Graphic board(s): standard Macintosh video card
☐ Mouse(s): standard Macintosh mouse
☐ Printer(s): ImageWriter and PostScript printers
☐ Minimum RAM: 512K
☐ Dual-floppy-disk system: yes
☐ Version: 2.0

Price and Vendor
☐ Industrial price: $595
☐ Academic price: $396
☐ Vendor: Cambridge Scientific Computing, Inc., P.O. Box 2123, Cambridge, MA 02238, (617)491-6862

 ChemIntosh

ChemIntosh is an Apple Macintosh desk accessory program that allows the user to call up the program from a Macintosh application like a word processor. The program has palette tools for placing symbol strings, drawing 10 different bond types, 11 different rings, and four straight

ChemIntosh Output

arrows: reaction, equilibrium, resonance, and dashed. It can draw solid or dashed circles, ovals, and arcs. Arrows may be added to either or both end points of an arc. ChemIntosh can optionally draw rings or bonds by using fixed bond lengths and angles and can customize bond widths, lengths, and angles. The program can automatically align structures on a single reference line, a feature useful to graphic artists. The program has layout rulers and marks to show the current drawing position and has a reduced view of the entire document page that assists the user in page layout. Other features include the ability to rotate and flip structures (without atomic labels) about the x and y axes, to join two molecular fragments into a single object, to move atoms and bonds independently from the rest of the molecule, to scale structures disproportionately in the x and y directions, and to group unrelated structures into a single object and then to ungroup them. The output from ChemIntosh is easily integrated into any Macintosh word-processing package, such as Microsoft Word, via the clipboard.

Hardware
☐ PC(s): Macintosh
☐ Graphic board(s): standard Macintosh video card
☐ Mouse(s): standard Macintosh mouse
☐ Printer(s): ImageWriter and PostScript printers
☐ Minimum RAM: 512K
☐ Dual-floppy-disk system: yes
☐ Version: 1.41

Price and Vendor
☐ Industrial price: $295
☐ Academic price: $236
☐ Student price: $118
☐ Vendor: SoftShell Company, P.O. Box 632, Henrietta, NY 14467, (716)334-7150

 ## ChemPlate/Hopkins

ChemPlate/Hopkins contains chemical-structure templates that can be used to assemble chemical-structure drawings on the Apple Macintosh by using the Apple Macintosh MacDraw program. To create a structure, the user selects templates from the ChemPlate library by pointing to them and clicking the mouse button. These primitive objects can then be grouped together to form more-complex structures. Users can create additional template structures, either by combining the units supplied by ChemPlate into new stored templates or by using MacDraw to make new structures.

ChemPlate/Hopkins Output

The ChemPlate library contains over 200 objects, including two-dimensional, three-dimensional, edgewise, and perspective representations of various hydrocarbon rings; perspective representations of tetrasubstituted C=C double bonds and tetrahedral carbon centers; Newman projection skeletons; nine sets of bonds oriented in various directions; arrowheads for drawing curved-arrow mechanisms and equations; and various symbols. Hopkins is a 36-point font in the correct ratio for lettering the ChemPlate drawings. The font character set includes superscripts, subscripts, and commonly used Greek letters and symbols.

Hardware
- ☐ PC(s): Macintosh
- ☐ Graphic board(s): standard Macintosh video card
- ☐ Mouse(s): standard Macintosh mouse
- ☐ Printer(s): ImageWriter and PostScript printers
- ☐ Minimum RAM: 512K
- ☐ Dual-floppy-disk system: yes
- ☐ Version: 1.2

Price and Vendor
- ☐ Price: ChemPlate/Hopkins is public-domain software. Copies of these programs can be obtained by sending a blank, initialized 3½ Macintosh disk together with $3.00 U.S. for postage and handling ($5.00 outside of North America) to the vendor.
- ☐ Vendor: Reuben Jih-Ru Hwu, Department of Chemistry, The Johns Hopkins University, Charles and 34th Streets, Baltimore, MD 21218

 ChemText

ChemText is a comprehensive text editor designed specifically for the chemical and pharmaceutical industries. The strengths of this software are its graphics programs: the Sketchpad and the Molecule Editor. The

Molecule Editor contains all the tools and templates needed to assemble a molecule or reaction sequence: atoms, bonds, stereochemical bonds, isotopes, charge, and valences. The Molecule Editor will advise the user when a structure is chemically incorrect. Once a structure has been drawn, individual atoms and bonds can be moved and the structure can be scaled to a different size.

The Sketchpad allows the user to use freehand techniques and individual graphic elements, such as lines, arrows, arcs, boxes, and ellipses, in preparing block diagrams, flow sheets, chemical-engineering drawings, etc. ChemText can also import graphic files from other graphics packages that use a compatible Molfile format for further modification. Alternatively, illustrations of molecules and reactions may be retrieved from other Molecular Design Limited software: ChemBase, MACCS, or REACCS. All graphics can be directly imported into a working text document created by the ChemText word processor.

ChemText Output

Hardware
☐ PC(s): IBM PCs and compatible systems
☐ Graphic board(s): CGA, EGA, and Hercules Graphics Card
☐ Mouse(s): Microsoft Mouse and Mouse Systems
☐ Printer(s): PostScript printers, Epson FX80 series and compatible printers, HP ThinkJet, HP LaserJet series, and HP LaserJet-compatible systems
☐ Minimum RAM: 640K
☐ Dual-floppy-disk system: no
☐ Version: 1.2

Price and Vendor
☐ Industrial price: $1500
☐ Academic price: $500
☐ Vendor: Molecular Design Limited, 2132 Farallon Drive, San Leandro, CA 94577, (800)635-0064 or (415)895-1313

ChemWord

ChemWord is a computer program for drawing and editing chemical
structures and merging them with text produced by a word processor. The
program interacts best with Microsoft Word in which structures can be
displayed; moved around; combined with text above, below, and to the
side; and merged with the printed Microsoft Word document. With other
word processors that are supported by the program, structures are merged
with the text when the document is printed. Users construct atoms and
bonds by accessing special character sets supplied by the program and can
create a library of structure fragments for later recall into a structure.

ChemWord Output

Hardware
□ PC(s): IBM PCs and compatible systems
□ Graphic board(s): CGA, EGA, and Hercules Graphics Card
□ Mouse(s): Microsoft Mouse
□ Printer(s): Epson FX80 series and compatible systems and HP LaserJet
   series and compatible systems
□ Minimum RAM: 180K
□ Dual-floppy-disk system: yes
□ Version: 2.0

Price and Vendor
□ Industrial price: $285
□ Academic price: $243
□ Vendor: Laboratory Software Ltd., 2 Ivy Lane, Broughton, Aylesbury
   HP22 5AP, England, (44)296-431234

 ChiWriter

ChiWriter is a character-based word processor that has chemical-structure-
drawing capabilities. ChiWriter comes with many font types, which contain

the bond fragments, letters, numbers, and special characters for building simple molecular structures in a WYSIWYG mode. The user can create a molecular structure in half-line mode, which allows creation of a structure as a single entity. Also, the user can switch to an asynchronous mode by which the individual parts of a structure can be moved separately to make final adjustments.

ChiWriter has the Font Designer program that allows the user to edit two font characters at once, for example, bond fragments in a special orientation. The edited fonts can then be juxtaposed, if desired, to form a larger character (e.g., bond fragment) beyond the boundaries of either one separately.

ChiWriter Output

Hardware
□ PC(s): IBM PCs and compatible systems
□ Graphic board(s): CGA, EGA, and Hercules Graphics Card
□ Mouse(s): not applicable
□ Printer(s): Epson FX80 series and compatible systems and HP LaserJet series and compatible systems
□ Minimum RAM: 256K
□ Dual-floppy-disk system: yes
□ Version: 3.0

Price and Vendor
□ Industrial and academic price: $99.95 + $49.95 for chemistry font; optional international keyboard, screen, and printer drivers at extra charge from $19.95 to $59.95
□ Vendor: Horstmann Software Design Corporation, P.O. Box 5039, San Jose, CA 95150, (408)298-0828

2D-CHEMICAL STRUCTURES

2D-CHEMICAL STRUCTURES is a program module of the Visper-32 software series. This software is independent of the operating system of the host computer because, once installed, a new 32-bit "virtual computer" is created within the memory that controls the program. 2D-CHEMICAL STRUCTURES produces representations of molecules containing up to 256 atomic centers and 256 bonds by using an invisible grid (256 by 256 points) whose x and y spacings may be adjusted independently. Structures may have single, double, triple, aromatic, complex, and stereochemical bonds. A font set provides all the Greek letters and a wide range of chemical and mathematical symbols, in addition to the usual alphanumeric characters.

Display options include "zoom" and "fit-in-window" modes, and structural representations may take various forms: explicit display of all atoms, identification of heteroatoms only, or carbon numbering. The program will keep track of normal valences, introduce hydrogen atoms as necessary, and calculate molecular formulas. A separate DRAW+PLOT module must be used to output structures to a plotter. All documentation for 2D-CHEMICAL STRUCTURES is in German. An English version is expected by Fall, 1988.

2-D CHEMICAL STRUCTURES Output

Hardware
☐ PC(s): IBM PCs and compatible systems; Atari Amega ST, 1040, and 520 ST; and Macintosh 512, 512E, Plus, and SE
☐ Graphic board(s): CGA, EGA, VGA, Hercules Graphics Card, and standard Macintosh video card
☐ Mouse(s): Microsoft Mouse, Mouse Systems, and Logitech
☐ Printer(s): Epson FX80 series and compatible systems, HP LaserJet series and compatible systems, and PostScript printers
☐ Plotters: HP 7475A, BBC SE 284, Tektronix 4662, and C. Itoh 4800
☐ Minimum RAM: 512K
☐ Version: 2.0

Price and Vendor
- ☐ Industrial price: 2900 DM (includes 2D-CHEMICAL STRUCTURES, DRAW+PLOT, and BASIS software)
- ☐ Academic price: 1675 DM (includes 2D-CHEMICAL STRUCTURES, DRAW+PLOT, and BASIS software)
- ☐ Vendor: VCH Verlagsgesellschaft, Scientific Software Division, Postfach 1260/1280, D-6940 Weinheim, Federal Republic of Germany, (49)6201-606156, or William E. Russey, 3732 Cold Springs Road, Huntingdon, PA 16652, (814)643-6793

 DrawStructures and Organic Fonts

DrawStructures is a library of object-oriented clip art for the Macintosh. The library contains a variety of organic, biochemical, and other classes of chemical compounds. DrawStructures files are in the PICT format, which allows the user to use other Macintosh programs, such as MacDraw, MacDraft, or SuperPaint, to compose, edit, and resize molecular-structure drawings for inclusion into word-processing documents. The manual contains an index of structures and provides instruction about the effective use of drawing programs and object-oriented images. Organic Fonts is a set of fonts designed to produce simple molecular structures by using MacPaint or FullPaint.

DrawStructures and Organic Fonts Output

Hardware
- ☐ PC(s): Macintosh
- ☐ Graphic board(s): standard Macintosh video card
- ☐ Mouse(s): standard Macintosh mouse
- ☐ Printer(s): ImageWriter and PostScript printers
- ☐ Minimum RAM: 512K
- ☐ Dual-floppy-disk system: yes
- ☐ Version: 1.0 (DrawStructures) and 1.2 (Organic Fonts)

Price and Vendor
☐ Industrial price: DrawStructures, $79.95; Organic Fonts, $79.95
☐ Academic price: DrawStructures, $71.95; Organic Fonts, $71.95
☐ Vendor: Modern Graphics, P.O. Box 21366, Indianapolis, IN 46221-0366, (317)253-4316

 ## GIOS

GIOS is a chemical-structure-drawing and -management program for the Apple IIe and IIc. The software consists of five interacting programs, two of which allow the user to create molecular-structure diagrams and schemes. The Structure Editor program creates molecular structures on a user-variable hexagonal grid by using input from an option menu and a mouse or a graphics tablet pen. For example, in the Chain mode, double, triple, alpha, beta, stereochemically undefined, or aromatic bonds can be drawn to specified grid points. In the Heteroatom function, the user can change the default carbon atoms to other atoms or functional groups.

GIOS uses the Data program module to automatically calculate and store the molecular formula, molecular weight, isotope mass of the molecular ion peak, and elemental analysis of a structure. The Environment Editor allows the user to combine structures created in the Structure Editor into reaction schemes containing reaction arrows, frames, annotating text, and tables. GIOS supports dot matrix printers and plotters for printing output with the Plot program module.

GIOS Output

Hardware
☐ PC(s): Apple IIe and IIc
☐ Graphic board(s): standard Apple II video card
☐ Mouse(s): standard Apple II mouse
☐ Graphics tablet(s): standard Apple II graphics tablet
☐ Printer(s): Epson FX80 series and compatible systems
☐ Plotter(s): Watanabe and Hewlett Packard 7475A, 7550A, and 7440
☐ Minimum RAM: 128K
☐ Dual-floppy-disk system: yes
☐ Version: 3.5

Price and Vendor
□ Industrial and academic price: 1200 DM
□ Vendor: Georg Thieme Verlag, Lektorat Chemie/Pharmazie, Rudiger-strasse 14, D-7000 Stuttgart 30, Postfach 732, D-7000 Stuttgart 1, Federal Republic of Germany, (49)711-8931-0

 Molecular Presentation Graphics (MPG)

MPG is a drawing program that allows users to create chemical structures and reaction schemes. With MPG, structure templates are accessed through menus that include benzene rings, polygons with up to nine sides, and many bond types: single, double, triple, forward, wedge, dashed, etc. When chains of bonds are drawn from rings, the program places the bonds at the correct angles automatically. Lines, arcs, and arrows also can be placed freehand to create reaction schemes. Drawings created by MPG can be merged (either between lines or with wrap-around text) with common word-processing documents at print time.

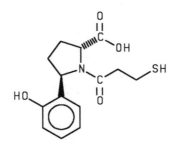

Molecular Presentation Graphics Output

Hardware
□ PC(s): IBM PCs and compatible systems, Wang PC, and Hewlett Packard 150
□ Graphic board(s): CGA, EGA, VGA, and Hercules Graphics Card
□ Mouse(s): Microsoft Mouse, Mouse Systems, and Logitech; cursor keys optional
□ Printer(s): Epson FX80 series and compatible systems and HP LaserJet series and compatible systems
□ Plotter: HPGL plotters
□ Minimum RAM: 256K
□ Dual-floppy-disk system: yes
□ Version: 4.1

Price and Vendor
- Industrial and academic price: $275
- Vendor: Hawk Scientific Systems, Inc., 170 Kinnelon Road, Suite 8, Kinnelon, NJ 07405, (201)838-6292

 PICASSO

PICASSO is a structure-drawing program for the IBM PC used with on-screen menus and a mouse. PICASSO diagrams may be drawn in either freehand or normalized mode by using an invisible grid system that ensures an even arrangement of bond lengths and angles. The user may also access templates, supplied by the program or user generated, to construct the structures. Bond types are chosen from a submenu and include bond specifications for multiple chains, rings, and stereochemical bonds. Atom types are also chosen from a submenu. PICASSO accounts for charge, delocalization, and mass attributes for each atom and creates connection tables that are used with PC-SABER for chemical database management. Higher quality diagrams of structures are printed by using a separate program module, DEGAS.

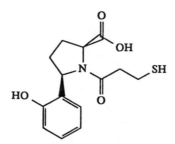

PICASSO Output

Hardware
- PC(s): IBM PCs and compatible systems
- Graphic board(s): CGA, EGA, and Hercules Graphics Card
- Mouse(s): Microsoft Mouse and Mouse Systems; cursor keys optional
- Printer(s): Epson FX80 series and compatible systems and Kyrosera laser printer
- Minimum RAM: 512K
- Dual-floppy-disk system: yes

Price and Vendor
- Industrial and academic price: PICASSO, £500; Degas, £200
- Vendor: Fraser Williams (Scientific Systems) Ltd., London House,

London Road South, Poynton, Cheshire SK12 1YP, England, (44)625-871126

 PsiGen

PsiGen is the main module in a suite of programs for creating, filing, and searching chemical-structure diagrams. PsiGen creates, edits, and displays chemical structures by using drop-down menus and a mouse. The user may use the templates of the program, freehand drawing, or the Feldmann notation (e.g., 66U6D5 for a steroid skeleton) to build molecular structures and can enter editing commands by menu "buttons" or keyboard characters. PsiGen performs valency checks and can calculate molecular formulas on the generated structures. Other PSIDOM program modules allow the user to integrate PsiGen-generated structures into word-processing files (PsiText), to build them into a database of structures, plot them on a HPGL-compatible plotter, and create electronic index cards (PsiBase).

PsiGen Output

Hardware
☐ PC(s): IBM PCs and compatible systems
☐ Graphic board(s): CGA, EGA, and Hercules Graphics Card
☐ Mouse(s): Microsoft Mouse and Mouse Systems; cursor keys optional
☐ Printer(s): Epson FX80 series and compatible systems and HP LaserJet series and compatible systems
☐ Plotter(s): HPGL plotters
☐ Minimum RAM: 512K
☐ Dual-floppy-disk system: yes
☐ Version: 5.1

Price and Vendor
☐ Industrial price: $495
☐ Academic price: $245

☐ Vendor: Hampden Data Services, Ltd., 167 Oxford Road, Cowley, Oxford OX4 2ES, England, (44)865-747250

 Spellbinder Scientific

Spellbinder Scientific is a technical word processor and office management system. The program has limited capabilities for creating chemical-structure diagrams, which can be keyed in as character-based fonts. The program is supplied with Greek-, mathematical-, and chemical-character sets and users may, with a character editor program, create their own chemical-character sets. Users may use macros to store and later recall multiline diagrams. Spellbinder Scientific allows the user to outline structural areas on screen so that a formula or chemical representation can be edited as a block. This feature allows the user to add to or delete from any structure and still maintain the horizontal and vertical alignments of each character.

Hardware
☐ PC(s): IBM PCs and compatible systems
☐ Graphic board(s): CGA, EGA, and Hercules Graphics Card
☐ Mouse(s): not applicable
☐ Printer(s): Epson FX80 series and compatible systems and HP LaserJet series and compatible systems
☐ Minimum RAM: 512K
☐ Dual-floppy-disk system: yes
☐ Version: 6.03

Price and Vendor
☐ Industrial price: $790
☐ Academic price: $711
☐ Vendor: Spellbinder Software Products, P.O. Box 1950, Davis, CA 95617, (916)758-3630
No output was provided by the vendor.

 T³

T^3 is a scientific word processor that works in the graphics mode to allow on-screen entry of chemical-structure diagrams. The program creates chemical structures through the use of special fonts that build the characters for horizontal, vertical, and diagonal bonds of a structure. T^3 has a menu-driven user interface with supporting help screens available. Over 1000 predefined macros are included with T^3 for mathematical symbols and chemical structures. Users may create their own macros or

T³ Output

use the predefined macros to retype structures with a few simple keystrokes. T³ can import graphics from other programs with selected file formats.

Hardware

☐ PC(s): IBM PCs and compatible systems
☐ Graphic board(s): CGA, EGA, VGA, and Hercules Graphics Card
☐ Mouse(s): not applicable
☐ Printer(s): Epson FX80 series and compatible systems and HP LaserJet series and compatible systems
☐ Minimum RAM: 640K
☐ Dual-floppy-disk system: no
☐ Version: 2.2

Price and Vendor

☐ Industrial price: $595
☐ Academic price: $476
☐ Vendor: TCI Software Research, Inc., 1190-B Foster Road, Las Cruces, NM 88001, (800)874-2383 or (505)522-4600

TechSet

TechSet is a scientific and technical typesetting program specifically designed for the HP LaserJet Plus printer with the B or F font cartridges. TechSet uses a mark-up syntax similar to a graphics programming language to create chemical-structure diagrams. Once a structure has been defined, it can be stored as a macro that can be concatenated with other macros to draw more-complicated structures. Because all the mark-up codes are in ASCII, TechSet does not require a graphics screen to create the structure diagrams.

TechSet Output

Hardware

☐ PC(s): IBM PCs and compatible systems
☐ Graphic board(s): not applicable
☐ Mouse(s): not applicable
☐ Printer(s): HP LaserJet series and compatible systems
☐ Minimum RAM: 384K
☐ Dual-floppy-disk system: yes
☐ Version: 2.0

Price and Vendor

☐ Industrial and academic price: $295
☐ Vendor: Software Development & Distribution Center, 1210 West Dayton
   Street, Madison, WI 53706, (800)262-6243 or (608)263-9484

 TechWriter

TechWriter is a WYSIWYG word processor that comes with a keyboard
template to help the user keep track of the sequence of functions needed
to access the many characters available in its alternate keyboards. All

TechWriter Output

Greek characters, several hundred technical and chemical characters, oversize characters (brackets, integral signs, etc.), diacritical marks, and sub- and superscripts appear on the screen in the same way they do in print. Standard text is entered in a text mode, and chemical structures are entered in a chemical-structure mode. Within the chemical-structure mode, framed areas (blocks) can be defined and moved as a unit to facilitate editing and positioning of bonds and atoms. An optional program called TechWriter Font Maker is available for the user to develop individual character sets for bond fragments.

Hardware
- □ PC(s): IBM PCs and compatible systems
- □ Graphic board(s): CGA, EGA, and Hercules Graphics Card
- □ Mouse(s): not applicable
- □ Printer(s): Epson FX80 series and compatible systems and HP LaserJet series and compatible systems
- □ Minimum RAM: 256K
- □ Dual-floppy-disk system: yes
- □ Version: 3.0

Price and Vendor
- □ Industrial price: $395
- □ Academic price: $295
- □ Vendor: CMI Software, 621 Main Street, Waltham, MA 02154, (617)899-7244

 The Egg

The Egg is a WYSIWYG word processor that has limited chemical-structure-drawing capabilities. The program has character fonts and a library of common chemical compounds for creating simple chemical structures. The chemical-structure fonts, which are literally bond fragments, are in an alternate keyboard that is accessed via the escape key followed by a number. When creating a chemical-structure diagram, the user can define multiple rows with up to 17 vertically spaced levels per row. The bond fragments are then entered one at a time to create the structure. The Egg also has a font-generating program that allows the user to create other needed bond fragments.

Hardware
- □ PC(s): IBM PCs and compatible systems
- □ Graphic board(s): CGA, EGA, and Hercules Graphics Card
- □ Mouse(s): not applicable

☐ Printer(s): Epson FX80 series and compatible systems and PostScript printers
☐ Minimum RAM: 256K
☐ Dual-floppy-disk system: yes
☐ Version: 4.2FF

Price and Vendor
☐ Industrial price: $495 + $125 for ChemLibrary Supplement
☐ Academic price: $396 + $112 for ChemLibrary Supplement
☐ Vendor: Peregrine Falcon Company, Suite B, 65 Koch Road, Corte Madera, CA 94925, (800)621-0851, extension 204, or (415)924-4663
No output was supplied by the vendor.

 WIMP

WIMP is a chemical-structure-drawing package developed by Howard Whitlock at the University of Wisconsin, Madison. Aldrich Chemical Company uses the program to produce *Aldrichimica Acta* and *Aldrich Catalog of Fine Chemicals*, as as well as other Aldrich publications.

WIMP treats the screen as a window onto a sheet of paper that accommodates large reaction schemes. Structures are drawn with a mouse or the keyboard numeric key pad, either freehand or by using geometric grids as a guide. The structures are stored as vector coordinates. In the editing mode, structures can be copied or called from disk, rotated, merged or docked with other structures, inverted, moved, or scaled. The program can create a variety of bond types, including single, multiple, hatched, thick, wedged, or jagged bonds, and can access a number of structure templates, including three- to nine-membered rings, cyclohexane conformations, and aromatic benzene, phenyl, and cyclopentadienyl rings.

WIMP Output

By using the "rubber band" feature, structures can be redrawn into new configurations. Atomic positions and captions can be printed with a variety of fonts and point sizes. WIMPlates, which contains a library of approximately 450 structure backbones and functional-group fragments, and Lab Gallery, an assortment of templates for laboratory glassware and apparatus, are also available for use with WIMP.

Hardware
- □ PC(s): IBM PCs and compatible systems
- □ Graphic board(s): CGA
- □ Mouse(s): Microsoft Mouse and Mouse Systems
- □ Printer(s): PostScript printers
- □ Plotter(s): HPGL plotters
- □ Minimum RAM: 512K
- □ Dual-floppy-disk system: yes
- □ Version: 6.0

Price and Vendor
- □ Industrial and academic price: WIMP, $250; WIMPlates, $95; Lab Gallery, $75.00
- □ Vendor: Aldrich Chemical Company, Inc., Technical Services Department, P.O. Box 355, Milwaukee, WI 53201, (800)231-8327 or (414)273-3850

WordMARC Composer+

WordMARC Composer+ is a WYSIWYG word processor that has limited chemical-structure-drawing capabilities. Bond fragments are accessed via an alternate keyboard by using a function key.

Hardware
- □ PC(s): IBM PCs and compatible systems
- □ Graphic board(s): CGA, EGA, and Hercules Graphics Card
- □ Mouse(s): not applicable
- □ Printer(s): HP LaserJet series and compatible systems and PostScript printers
- □ Minimum RAM: 512K
- □ Dual-floppy-disk system: yes
- □ Version: 6.1.1

Price and Vendor
- □ Industrial and academic price: $495

□ Vendor: MARC Software International, Inc., 260 Sheridan Avenue, Palo Alto, CA 94306, (415)326-1971

No output was supplied by the vendor.

Reference

1. Bentley, J.; Jelinski, L.; Kernighan, B. *Comput. Chem.* **1987**, *11*, 281.

CHAPTER THREE

Graphics Terminal Emulation Software

Wendy A. Warr

Communication software allows a user to move information to and from a PC and a remote computer. With this software, commands from the PC are sent out via a modem or LAN (local area network), and information from the remote computer is displayed on the PC screen or downloaded into its diskettes. For some online files, videotext systems, and electronic bulletin boards, the displayed information contains graphics data. Graphics terminal emulation is a type of communication software and associated hardware that enables a PC to simulate the characteristics of the graphics terminal supported by the remote hosts. (For more information on the use of microcomputers for online searching, *see* reference 1.) One complication of the use of graphics emulation software is the compatibility between the keyboard and display attributes of the PC and the remote hosts.

This chapter focuses on software that emulates graphics terminals, such as those needed to access chemical-structure databases. MACCS is an example of an inhouse database system that contains chemical structures, and DARC/Questel, CAS ONLINE, and the Chemical Information System (CIS) are examples of public online database systems.

Use of Graphics Terminal Emulation

Until communication software for PCs became available, expensive graphics terminals were needed to access remote graphics databases. The availability of terminal emulation software permitted the use of relatively low-cost PCs for online graphics searching. The communication aspect of the software also allowed the automation of the log-on and search process. Further development in emulation software led to the offline building of structure graphics and subsequent uploading to a host computer, as well as the capture (downloading) of retrieved records.

Front-end software allows the user to formulate searches offline and then to upload them after logging on to a desired system. This software also assists the user in formulating a search and capturing data. Front-end software for structure search is normally limited to a single data system: STN Express for CAS ONLINE and DARC CHEMLINK for DARC/ Questel. MOLKICK serves as a front end for the Beilstein database on several online systems. Single-system front ends are discussed in Chapter 6 as special-application software. The eight software packages described in this chapter are broad-utility communication–emulation programs that can be used to access a wide variety of inhouse and online structure database systems.

Graphics Terminal Emulation Programs

The programs for graphics terminal emulation that are discussed in this chapter are the following.
- ChemTalk (Molecular Design Limited)
- EMU-TEK (FTG Data Systems)
- PC-PLOT-IV Plus (Microplot Systems)
- Reflection 7 Plus (Walker Richer and Quinn)
- SmarTerm 240 (Persoft)
- TextTerm + Graphics (Mesa Graphics)
- TGRAF-05, -07, and -15 (Grafpoint)
- VersaTerm and VersaTerm-PRO (Peripherals Computers and Supplies)
  The following are related software discussed elsewhere in this book.
- ChemText (*see* Chapter 2)
- ChemBase (*see* Chapter 4)
- DARC CHEMLINK, MOLKICK, and STN Express (*see* Chapter 6)

An additional graphics terminal emulation software that is available but was not included in this survey is G'POT from Kinokuniya Company, Inc. G'POT is a graphics emulation–communication software for NEC–DOS PCs. Interested readers should direct their inquiries to Kinokuya Information Retrieval Services, Dept. 38-1, Sakuragaoka 5-Chrome, Setagaya-ku, Tokyo (156), Japan.

Product Descriptions

 ### ChemTalk

ChemTalk is a communication–emulation program that can be used to share files between the ChemBase and ChemText PC programs of Molecular Design Limited (MDL) and mainframe databases. ChemHost is required on the mainframe if the program is used to access the MACCS-II or REACCS files of MDL. In general, the program allows direct access to inhouse and online graphic systems. ChemTalk emulates VT100 and VT640 and Tektronix 4010 and 4105 terminals.

Hardware
- PC(s): IBM PCs and compatible systems
- Graphics board(s): CGA, EGA, VGA, and Hercules Graphics Card
- Mouse(s): Microsoft Mouse and Mouse Systems
- Printer(s): Epson FX80 series and compatible systems, HP ThinkJet, HP LaserJet series and compatible systems, and PostScript printers
- Minimum RAM: 640K
- Dual-floppy-disk system: no
- Version: 1.2

Price and Vendor
- Industrial price: $1000
- Academic price: $450
- Vendor: Molecular Design Limited, 2132 Farallon Drive, San Leandro, CA 94577, (800)635-0064 or (415)895-1313

 ### EMU-TEK

EMU-TEK enables an IBM PC or compatible system to function as a Tektronix graphics terminal. EMU-TEK Seven Plus emulates a Tektronix 4107, and EMU-TEK Five Plus emulates a Tektronix 4105. EMU-TEK Level 2 emulates Tektronix 4010 and 4014 and adds an emulation for VT640 from Digital Engineering. EMU-TEK Level 1 emulates a Tektronix 4010 terminal. EMU-TEK terminal emulators are compatible with chemical-structure software such as CAS ONLINE, ChemQuest, CHEM-X, DARC, MACCS, REACCS, and SYBYL.

Hardware
- PC(s): IBM PCs and compatible systems
- Graphics board(s): CGA, EGA, VGA, and Hercules Graphics Card for

EMU-TEK Levels 1 and 2; and EGA and VGA for EMU-TEK Five Plus and Seven Plus
- ☐ Mouse(s): Microsoft Mouse, Mouse Systems, and AT&T
- ☐ Printer(s): Epson FX80 series and compatible systems and HP LaserJet series and compatible systems
- ☐ Plotter(s): NEC P5XL and HP7475
- ☐ Minimum RAM: 256K
- ☐ Dual-floppy-disk system: yes
- ☐ Version: 1.01E (Levels 1 and 2) and 1.32 (Five Plus and Seven Plus)

Price and Vendor
- ☐ Industrial price: EMU-TEK Level 1, $95; EMU-TEK Level 2, $295; EMU-TEK Five Plus, $495; EMU-TEK Seven Plus, $695
- ☐ Academic price: Call for quote.
- ☐ Vendor: FTG Data Systems, 10801 Dale Street, Suite J-2, Stanton, CA 90680, (714)995-3900

PC-PLOT-IV Plus

PC-PLOT-IV Plus is a terminal emulation package that provides support for the DEC VT100 and VT200 and Tektronix 4010, 4014, 4027, and 4105 terminals. The program includes autodial from a script file, graphics screen print, Xmodem and Kermit file transfers, keyboard redefinition, and programmable function keys. The software can be used to access CAS ONLINE, DARC/Questel, MACCS, REACCS, and other chemical-graphics systems.

Hardware
- ☐ PC(s): IBM PCs and compatible systems
- ☐ Graphics board(s): CGA, EGA, and Hercules Graphics Card
- ☐ Mouse(s): Microsoft Mouse and Mouse Systems
- ☐ Printer(s): Epson FX80 series and compatible systems and HP LaserJet series and compatible systems
- ☐ Minimum RAM: 256K
- ☐ Dual-floppy-disk system: yes
- ☐ Version: 4.20

Price and Vendor
- ☐ Industrial and academic price: $195
- ☐ Vendor: Microplot Systems, 659-H Park Meadow Road, Westerville, OH 43081, (614)882-4786

Reflection 7 Plus

Reflection 7 Plus provides the IBM PC and HP Vectra computers with complete emulation for HP2627A color, HP2623A monochrome, Tektronix 4010, and VT102 terminals. The program includes file transfer, command language, backup, and LAN support, along with the ability to display graphs with up to eight colors. Reflection 7 Plus provides a screen resolution of 640 × 350 pixels. Graphics features include "poly-gonfil", separate alphanumeric and graphics cursors and memories, and rubber band line. Graphics text can be created in eight sizes, either upright or slanted. Plotters can be attached via serial port in either eavesdrop or pass-through mode. Bit-mapped graphics printing is available on supported printers.

Hardware
□ PC(s): IBM PCs and compatible systems
□ Graphics board(s): CGA, EGA, and VGA
□ Mouse(s): Microsoft Mouse and Mouse Systems
□ Printer(s): Epson FX80 series and compatible systems, HP LaserJet series and compatible systems, and HP ThinkJet
□ Minimum RAM: 512K
□ Dual-floppy-disk system: yes
□ Version: 3.0

Price and Vendor
□ Industrial and academic price: $449
□ Vendor: Walker Richer and Quinn Inc., 2825 Eastlake Avenue East, Seattle, WA 98102, (206)324-0350

 SmarTerm 240

SmarTerm 240 allows an IBM PC to emulate DEC VT241 color, DEC VT240, DEC VT125, and Tektronix 4014 graphics terminals. SmarTerm 240 also provides text emulation for the DEC VT220, VT100, VT102, and VT52 text terminals, as well as TTY mode for access to popular time-sharing services. The program provides three error-free file-transfer protocols: Kermit, Xmodem, and Persoft's proprietary PDIP protocol.

Hardware
□ PC(s): IBM PCs and compatible systems
□ Graphics board(s): CGA, EGA, and Hercules Graphics Card
□ Mouse(s): not applicable

☐ Printer(s): Epson FX80 series and compatible systems and HP Laserjet series and compatible systems
☐ Minimum RAM: 512K
☐ Dual-floppy-disk system: yes
☐ Version: 2.0

Price and Vendor
☐ Industrial and academic price: $345
☐ Vendor: Persoft Inc., 465 Science Drive, Madison, WI 53711, (608)273-6000

 ## TextTerm + Graphics

TextTerm + Graphics provides emulation of the Tektronix 4014, 4012, 4006, and 4106 and the DEC VT100 terminals for the Macintosh PCs. The program supports the Xmodem and Kermit transfer protocols.

Hardware
☐ PC(s): Macintosh
☐ Graphics board(s): standard Macintosh video card
☐ Mouse(s): standard Macintosh mouse
☐ Printer(s): ImageWriter and PostScript printers
☐ Minimum RAM: 512K
☐ Dual-floppy-disk system: yes
☐ Version: 1.02

Price and Vendor
☐ Industrial and academic price: $195
☐ Vendor: Mesa Graphics Inc., P.O. Box 600, Los Alamos, NM 87544, (505)672-1998

TGRAF

TGRAF-05, -07, -15, and -15LR are graphics terminal emulation packages for the IBM PC and compatible systems. TGRAF-05 accesses both alphanumeric and graphics application software running on larger computers. As an alphanumeric terminal, TGRAF-05 enables the PC to emulate a DEC VT100 terminal. In graphics mode, TGRAF-05 provides Tektronix 4105 emulation. The functions of TGRAF-05 include points, vectors, filled panels, graphic text, Tektronix 4010 and 4014 modes, VT52 and VT100 modes, file transfer, local storage and retrieval of graphics images, and keyboard reprogrammability. TGRAF-07 provides all the functions of TGRAF-05 plus additional functions found on the Tektronix 4107, 4109, and 4200 series graphics terminals. TGRAF-15 and TGRAF-15LR (low resolution) combine the features of TGRAF-05 and TGRAF-07 with additional features of the Tektronix 4115 graphics terminal.

Additional supporting modules include TNET and TPORT. Both programs are full-functioned graphics terminals installed as device drivers on the PC. TNET enables a host and a PC to communicate via a local area network. TPORT allows mainframe- and minicomputer-based software to be ported to the PC.

Hardware
□ PC(s): IBM PCs and compatible systems, Macintosh II, and Silicon Graphics workstation
□ Graphics board(s): CGA, EGA, and Hercules Graphics Card
□ Mouse(s): Microsoft Mouse and Mouse Systems
□ Graphics tablet(s): Summagraphics
□ Printer(s): Epson FX80 series and compatible systems and HP LaserJet series and compatible systems
□ Minimum RAM: TGRAF-05 and TNET, 128K; TGRAF-07, TGRAF-15, and TGRAF-15LR and TPORT, 256K
□ Dual-floppy-disk system: yes
□ Version: 2.2

Price and Vendor
□ Industrial price: TGRAF-05, $395; TGRAF-07, $995; TGRAF-15LR, $1495; TGRAF-15HR, $1995; TNET and TPORT, $395–1995
□ Academic price: Call for quote.
□ Vendor: Grafpoint, 1485 Saratoga Avenue, San Jose, CA 95129-4934, (408)446-1919 or (800)426-2230

 ## VersaTerm and VersaTerm-PRO

VersaTerm is a text-and-graphics communication–terminal emulator suitable for file transfer, information services, and minicomputer-to-mainframe access. The terminals it supports are DEC VT100; Tektronix 4010, 4012, and 4014; and Data General D200. Graphics data is turned into a bit-map when received from the host, a process allowing MacPaint-type graphics storage. File-transfer capability is compatible with Xmodem and Kermit protocols. VersaTerm products have automatic macro capabilities that streamline communications. All autodial modems are supported. The current version of VersaTerm supports both the new Macintosh SE and

Macintosh II computers. VersaTerm supports multitasking background operation of file transfers, printing, and terminal sessions with Apple's new MultiFinder.

VersaTerm-PRO is a text-and-color graphics communication–terminal emulator that includes all the features that are found in VersaTerm: DEC VT100; Tektronix 4010, 4012, and 4014; and Data General D200 emulations. In addition, VersaTerm-PRO provides complete Tektronix 4105 color-graphics emulation. Programs such as MACCS and REACCS for chemical design, Intergraph for CAD/CAM, and SAS/Graph for presentation graphics are examples of mainframe graphics programs that can be accessed by Macintosh computers equipped with VersaTerm series software. VersaTerm-PRO maintains the full high-resolution vector image of the graphic in memory and allows zoom and pan with either the 4105 or 4014 emulation. All autodial modems are supported.

Hardware
- PC(s): Macintosh II and Macintosh SE
- Graphics board(s): standard Macintosh video card
- Mouse(s): standard Macintosh mouse
- Printer(s): ImageWriter and PostScript printers
- Minimum RAM: 512K
- Dual-floppy-disk system: yes
- Version: 3.2 (VersaTerm and VersaTerm-PRO)

Price and Vendor
- Industrial and academic price: VersaTerm, $99; VersaTerm-PRO, $295
- Vendor: Peripherals Computers and Supplies Inc., 2457 Perkiomen Avenue, Reading, PA 19606, (215)779-0476

Reference

1. *Microcomputers for the On-line Searcher;* Alberico, R., Ed.; Meckler Publishing: Westport, CT, 1987.

CHAPTER FOUR

Structure Management Software

Daniel E. Meyer

S tructure management software differs from graphics emulation and structure-drawing software in that it includes the capability to retrieve data via exact structure and substructure searches. With this software, the structure graphic data is stored in a connection table format rather than as a vector or bit-mapped file. Some of the programs may also store, search for, and retrieve related numeric and text data.

Uses of Structure Management Software

Structure records in emulation programs are stored as vector files and can only be recalled by identification name or number. Structure-drawing programs may store compounds as vector files, in which case graphic retrieval is not possible, or as connection tables, in which case structure retrieval would be possible if structure search routines were available with the software. It is likely that some of the drawing programs which store compound information in connection table format will be expanded to include structure retrieval in the future. One of the programs discussed in this chapter, PsiBase/PsiGen, was first released as a structure-drawing package (PsiGen), and then a structure retrieval module (PsiBase) was added later.

Structure management software provides users the opportunity to build personal databases of structures and related data. Each program included

1538–3/88/0045/$06.00/0 © 1988 American Chemical Society

in this chapter allows the graphic input of a compound, the registration of the compound into a database, and the retrieval of records via structure and substructure searches. With these features, researchers can create personal databases of compounds reported in the literature, structures identified in their research, or inventories of chemicals available in their company. Information about a specific compound or group of compounds can be easily gathered by conducting the appropriate structure search.

Requirements and Features of Structure Management Software

All the programs discussed in this chapter are written for IBM-compatible PCs. One program, CHEMDATA, also runs on the Macintosh SE. The size of the database that the programs can create is limited by the storage capacity of the user's hard disk and by the program's search speed. A personal database of 2000–15,000 compounds is the maximum realistic size for these products. TREE (HTSS) is capable of rapid search speeds (approximately 20–25 s) for a large file of 15,000 compounds.

The hardware requirements vary greatly for these programs. Of the seven programs included in this chapter, only two, ChemBase and CHEMDATA, require a mouse for input. The other five programs can use either a mouse or keyboard cursor keys. The use of a hard disk and a minimum memory of 512K is either recommended or required for four of the products: ChemBase, PC-SABRE/PICASSO, PsiBase/PsiGen, and TREE (HTSS). ChemFile II and ChemSmart both work well in a dual-floppy-disk system with a memory of 256K. CHEMDATA will run on a dual-floppy-disk system but requires a minimum memory of 512K.

ChemBase, PC-SABRE/PICASSO, and PsiBase/PsiGen use the same Molecule Editor as their related structure-drawing modules and, therefore, can give presentation-quality graphics output (*see* Chapter 2). All the programs have clean graphics screen displays so that the atoms and bond types are distinctly visible. All the programs can depict a variety of bond types including, in- and out-of-plane bonds for stereochemical designation.

An important feature in this group of software is the inclusion of chemical expertise. All the programs, except CHEMDATA and ChemFile II, check for proper valence when a new compound is entered. Chem-Base, CHEMDATA, and PsiBase/PsiGen calculate the molecular formula. ChemBase and CHEMDATA also calculate the molecular weight. The ability to search the related numeric and text data varies greatly among these programs. ChemBase has the greatest search capability and allows the search of all data elements, including range search of numeric data.

ChemFile II also allows text retrieval and range search of specific fields. ChemSmart only allows the search of the fields for compound name and molecular formula. TREE (HTSS) and PsiBase/PsiGen provide links to other text management software for data retrieval. PC-SABRE/PICASSO allows text storage and the display of data linked to retrieved structures.

One distinguishing feature of ChemBase is that, in addition to being a structure management program, it can build, store, and retrieve reaction schemes. ChemBase can search specifically for the product or reactant of a reaction. Two other programs allow reaction schemes to be displayed but do not provide true reaction searches.

Structure management programs offer the possibility of developing links to other programs and allow the direct transfer of data between programs. Several additional programs are being developed that will expand the use of these programs so that a more integrated chemical workstation can be provided to researchers. This integrated workstation environment will allow the local storage of structural records and the transfer of the information to online and on-site data systems or to modeling and drawing programs.

Product Selection

I have reviewed all the following programs, except PC-SABRE/PICASSO and CHEMDATA, and have included these programs in papers presented at national meetings of the American Chemical Society. The data for the product description section were provided by the software producers, but I have also reviewed these data during use of the programs.

Structure Management Programs

The programs for structure management that are discussed in this chapter are the following.
- ChemBase (Molecular Design Limited)
- CHEMDATA (VCH Verlagsgesellschaft)
- ChemFile II (Queue)
- ChemSmart (ISI Software)
- PC-SABRE/PICASSO (Fraser Williams)
- PsiBase/PsiGen (Hampden Data Services)
- TREE (HTSS) (Technical Database Services)

The following are related software that are discussed elsewhere in this book.

- ChemText, 2D-CHEMICAL STRUCTURES, PICASSO, and PsiGen (*see* Chapter 2)
- ChemTalk (*see* Chapter 3)
- CASKit-1 and -2, PC-MARKOUT, and TopKat (*see* Chapter 6)

An additional structure management program that is available but was not included in this survey is TopDog, which is available from Health Designs, Inc., 183 East Main Street, Rochester, NY 14604, (716)546-1464.

Product Descriptions

 ## ChemBase

ChemBase is a database management system for creating, storing, and retrieving chemical structures and reactions. Users may search and sort compounds and reactions on the basis of structure, substructure, numerical, or text data. Databases may be downloaded from or uploaded to MACCS-II or REACCS via ChemTalk and ChemHost. Data fields can be customized for specific needs. ChemBase is fully integrated with Chem-Talk and ChemText. An initial database of structures and reactions is provided with the program, as well as a large collection of templates. ChemBase requires a mouse for graphics input and works in a hard-disk environment. ChemBase provides system expertise by checking for proper valence, automatically generating molecular formula and molecular weight, and highlighting atomic overlap. ChemBase permits direct building of reaction databases and reaction-specific searches.

Hardware
☐ PC(s): IBM PCs and compatible systems
☐ Graphic board(s): CGA, EGA, and Hercules Graphics Card
☐ Mouse(s): Microsoft Mouse and Mouse Systems
☐ Printer(s): PostScript printers, Epson FX80 series and compatible systems, HP ThinkJet, and HP LaserJet series and compatible systems
☐ Minimum RAM: 640K
☐ Dual-floppy-disk system: no
☐ Version: 1.2

Price and Vendor
☐ Industrial price: $3500
☐ Academic price: $975
☐ Vendor: Molecular Design Limited, 2132 Farallon Drive, San Leandro, CA 94577, (800)635-0064 or (415)895-1313

CHEMDATA

CHEMDATA is part of the Visper-32 software series from VCH Verlagsge-sellschaft. CHEMDATA allows graphic input of structures and related text information, such as compound name, molecular formula, molecular weight, and comments. Structures may be retrieved via structure and substructure searches, and different precision levels of searches can be conducted. A rapid search can be used to create a subset of candidate structures, and this subset can be refined by a precision search. Text data are searchable, and a range of search option exists for numerical fields. A second version of the program, CHEMDATA II, is available that also stores and searches spectral data.

Hardware
☐ PC(s): IBM PCs and compatible systems and Macintosh SE
☐ Graphic board(s): CGA, EGA, and Hercules Graphics Card
☐ Mouse(s): Microsoft Mouse and Mouse Systems
☐ Printer(s): Epson FX80 series and compatible systems, HP LaserJet series and compatible systems, and PostScript printers
☐ Minimum RAM: 512K
☐ Dual-floppy-disk system: yes
☐ Version: 2.0

Price and Vendor
☐ Industrial and academic price: 2200 DM
☐ Vendor: VCH Verlagsgesellschaft, Scientific Software Division, Postfach 1260/1280, D-6940 Weinheim, Federal Republic of Germany, (49)6201-606156, or William E. Russey, 3732 Cold Springs Road, Huntingdon, PA 16652, (814)643-6793

 ChemFile II

ChemFile II is a structure management program that is marketed by Queue, Inc. The methods for entering structures are straightforward, and the program generates graphic output of relatively good quality. Stereo-chemical bonds have been added in this updated version, as well as greater text storage and search capabilities. An important feature is a report generator that allows text data to be sorted and displayed in table format. ChemFile II works on a dual-floppy-disk 256K-memory system, and when a file is being searched, the program asks whether the user wants to continue the search on another diskette. This feature allows user files to grow beyond the storage capacity of a single diskette.

Hardware
- ☐ PC(s): IBM PCs and compatible systems
- ☐ Graphic board(s): EGA, CGA, and Hercules Graphics Card
- ☐ Mouse(s): Microsoft Mouse and Mouse Systems; cursor keys optional
- ☐ Printer(s): Epson FX80 series and compatible systems
- ☐ Minimum RAM: 256K
- ☐ Dual-floppy-disk system: yes
- ☐ Version: 1.0

Price and Vendor
- ☐ Industrial and academic price: $250
- ☐ Vendor: Queue, Inc., 562 Boston Avenue, Bridgeport, CT 06610, (800)232-2224 or (203)335-0908

ChemSmart

ChemSmart allows the building, storage, and retrieval of chemical structures. ChemSmart provides menus to assist in the input of the structural diagrams and stores the compound information with an identification number and related chemical data. A note card feature stores related information such as chemical name, molecular formula, molecular weight, physical and chemical properties, and bibliographic data. Only the fields for the molecular formula, compound name, and registration number are searchable. An initial library of 250 compounds is provided with the program diskette. The program works on a 256K-memory PC, and the use of a mouse is optional. ChemSmart checks each new entry for correct valence and provides templates for rapid input of structures.

Hardware
- ☐ PC(s): IBM PCs and compatible systems
- ☐ Graphic board(s): CGA, EGA, and Hercules Graphics Card
- ☐ Mouse(s): Microsoft Mouse and Mouse Systems; cursor keys optional
- ☐ Printer(s): Epson FX80 series and compatible systems
- ☐ Minimum RAM: 256K
- ☐ Dual-floppy-disk system: yes
- ☐ Version: 1.0

Price and Vendor
- ☐ Industrial and academic price: $199
- ☐ Vendor: ISI Software, 3501 Market Street, Philadelphia, PA 19104, (800)523-1850 or (215)386-0100

PC-SABRE/PICASSO

PC-SABRE provides facilities for input, registration, storage, search, retrieval, and display of chemical structures and associated data. Input is either graphic via PICASSO or from WLN (Wiswesser line notation) via DARING and REWARD. Search structures are entered graphically via PICASSO, which gives graphical input facilities for chemical structures through mouse-controlled menus. No keyboard entry is required during input. The display module, DEGAS, can be used to prepare presentation-quality graphics output. The PC-SABRE package includes PICASSO and DEGAS modules.

Hardware
□ PC(s): IBM PCs and compatible systems
□ Graphic board(s): CGA, EGA, and Hercules Graphics Card
□ Mouse(s): Microsoft Mouse and Mouse Systems; cursor keys optional
□ Printer(s): Epson FX80 series and compatible systems
□ Minimum RAM: 512K
□ Dual-floppy-disk system: no
□ Version: 1.0

Price and Vendor
□ Industrial and academic price: £1520
□ Vendor: Fraser Williams (Scientific Systems) Ltd., London House, London Road South, Poynton, Cheshire SK12 1YP, England, (44)625-871126

 ## PsiBase/PsiGen

PsiGen allows graphic input of structural diagrams (*see* Chapter 2). PsiBase enables the user to search databases built in PsiGen by either exact structure or substructure queries. Other modules are available that link text to structural information (PsiText) or generate reports containing structures from the database (PsiRep). PsiBase/PsiGen works best with a hard disk and a minimum memory of 512K. Either a mouse or cursor controls may be used for input, and PsiBase/PsiGen has a broad range of input techniques, such as ring and chain commands, as well as Feldmann notation. PsiBase/PsiGen automatically generates the molecular formula and verifies the correct valence for each new compound.

Hardware
□ PC(s): IBM PCs and compatible systems
□ Graphic Board(s): CGA, EGA, and Hercules Graphics Card

☐ Mouse(s): Microsoft Mouse and Mouse Systems; cursor keys optional
☐ Printer(s): Epson FX80 and compatible systems and HP LaserJet series
and compatible systems
☐ Minimum RAM: 512K
☐ Dual-floppy-disk system: yes (but not recommended)
☐ Version: 5.1

Price and Vendor
☐ Industrial price: $990
☐ Academic price: $495
☐ Vendor: Hampden Data Services, Ltd., 167 Oxford Road, Cowley,
Oxford OX4 2ES, England, (44)865-747250

TREE (HTSS)

TREE is the U.S. version of the Hierarchical Tree Substructure Software
(HTSS). TREE can retrieve matching chemical structures and substruc-
tures from large PC-based databases. TREE matches the input query
structure with the retrieved results by using a side-by-side window display.
The program finds the most similar structures if an identical match can
not be found and displays Markush structures from chemical fragments in
multiple windows. TREE can be used with standard word-processing
software to add chemical structures to text. It can also be used with dBASE
III+ and other database packages. An initial structure database library is
provided with the program.

Hardware
☐ PC(s): IBM PCs and compatible systems
☐ Graphic board(s): CGA, EGA, and Hercules Graphics Card
☐ Mouse(s): Microsoft Mouse and Mouse Systems; cursor keys optional
☐ Printer(s): Epson FX80 series and compatible systems
☐ Minimum RAM: 640K
☐ Dual-floppy-disk system: no
☐ Version: 3.11

Price and Vendor
☐ Industrial price: $499 without Markush or $899 with full Markush
☐ Academic price: $400 without Markush or $720 with full Markush
☐ Vendor: Technical Database Services, Inc., 10 Columbus Circle, Suite
2300, New York, NY 10019, (212)245-0044

CHAPTER FIVE

Software for Three-Dimensional Molecular Graphics and Modeling

Cyrelle K. Gerson

In describing concepts of molecular structures, reactions, and interactions, chemists use three-dimensional (3-D) images and models of atoms and molecules. Most chemists working in the field today were educated when a variety of wooden, metal, and plastic models were used to illustrate how a molecule "looked", to speculate on the feasibility of its existence, and even to predict some of its physical properties. While organic and inorganic chemists were using these crude approaches to represent molecules, theoretical chemists were developing sophisticated, quantitative theories, calculations, and computer graphics to simulate and predict molecular properties more accurately. Although the truly high-powered modeling programs are still limited to powerful mainframe and minicomputers, the development of programs for 3-D graphics and modeling for PCs has put some of the power of computer-aided modeling within the reach of most chemists.

The PC programs discussed in this chapter fall into two categories: (1) 3-D-drawing programs and (2) molecular-modeling programs. The drawing programs may also serve as front ends to more-powerful mainframe or minicomputer modeling systems.

Basically, the 3-D-drawing programs are limited to drawing and displaying structures and manipulating the displays. The structural data are sometimes stored in a format that is compatible with other modeling or drawing programs, a feature that allows some of these programs to

1538–3/88/0053/$06.00/0 © 1988 American Chemical Society

serve as front ends to mainframe or minicomputer systems. The front-end programs may be accompanied by communication software that allows the user to easily transfer data from the PC to the larger computer.

In addition to drawing and displaying structures, the molecular-modeling programs can calculate and predict a variety of physicochemical features of the input molecules. Many of these programs can also create data files that are compatible with modeling programs on larger computers.

Program Features

The key features of these programs that potential users can compare fall into three categories: methods of structure input, types of structure display and printout, and kinds of physicochemical features that can be calculated.

Most of the programs (both structure drawing and modeling) allow freehand structure input with a mouse or input from data tables such as crystallographic data or internal Cartesian coordinates. The programs that allow freehand input include internal definitions of the lengths and angles of many bond types between atoms. Some programs are restricted to simple organic molecules, and others can include virtually every element in the periodic table.

Typical structure display options include stick, ball-and-stick, and space-filling displays. Some of the modeling programs can also generate an ORTEP (Oak Ridge thermal ellipsoid plot) display. Several examples are shown on pages 56 and 57.

All of the programs allow 3-D structure rotation in at least one of the display modes. The more powerful programs allow real-time rotation, which gives a feeling of animation. Several of the drawing programs achieve this motion by allowing the user to store a series of views of the structure that can be programmed to replay in rapid sequence. With all of the modeling programs and some of the drawing programs, the user can enlarge or reduce the display or zoom in on a part of the molecule.

All of the modeling programs and almost all of the drawing programs support some type of hard-copy output. The quality of output devices supported varies greatly from high-quality plotters and laser printers to draft-quality dot matrix printers. The fact that the vast majority of these programs support some type of printed output indicates that most of the program developers and vendors recognize the need of chemists to communicate on paper. However, these programs generally are oriented to the achievement of excellent graphics display on the computer CRT rather than to the attainment of the best quality hard copy.

Distinguishing Features of Modeling Programs

One feature that distinguishes the modeling programs from the drawing programs described in this chapter and in Chapter 1 is the ability of the modeling programs to calculate several types of physical parameters. These calculations are at the heart of the theoretical chemist's simulation and prediction of molecular properties and, in the long run, amount to a great deal more than mere 3-D graphics display.

All of the modeling programs include some type of energy minimization function, either through molecular orbital or molecular mechanics calculations or both. To perform these CPU-intensive computations the user generally inputs the structure, views it in the graphics display mode, exits from the display mode by using a menu or series of commands to allow the minimization program to do its work, and then returns to the graphics display mode to view the molecular conformation with the lowest computed energy. Most of the modeling programs include their own unique minimization programs in addition to supporting programs or including such public-domain programs as MM2 and MNDO. Some programs also allow conformational analysis, that is, computation of the relative energy of each conformation of a molecule. CAMSEQ/M and Microchem—Organic Unit are two programs that do calculations of conformational analysis.

Often the goal of molecular modeling is to determine whether the shape of all or part of the molecule is the same as that of a known active compound. Several of the modeling programs allow graphical docking or geometric comparison of two structures to help achieve this goal. None of the PC-based modeling programs allow the user to analyze the interactions of a molecule with a receptor site. This analytic capability is included in some minicomputer and mainframe systems and is particularly useful in computer-aided drug design.

The Microchem series of programs includes a unit, GAP (Group Additive Properties), that is not a molecular-graphics program but is included in this chapter because it is a molecular-modeling tool. For example, one feature of GAP is that it can be used to predict the glass transition temperatures of homo- and copolymers. GAP also includes programs in its Drug Design module that establishes a database of empirically derived properties. GAP can search the database for substituents with one or more properties that approximate those of another substituent being studied by the user.

Limitations

Generally, the PC modeling programs are limited in the variety of computations they can make, their speed of operation, and the resolution of their displays compared with programs available on larger, more-powerful computers. Most of the PC programs require a math coprocessor and function best with a high-resolution graphics display (e.g., an IBM Enhanced Graphics Adaptor). Their performance, therefore, cannot be compared with that of a sophisticated program running on a more powerful computer (e.g., a VAX or MicroVAX) with higher resolution display (e.g., an Evans & Sutherland graphics terminal).

Despite their limitations, all of these programs add significantly to the ability of chemists to understand molecular properties, to teach students the concepts of molecular structure and reactivity, and to design new and useful compounds. Their reasonable cost, user friendliness, and availability make them accessible to more chemists than the expensive mainframe programs are. As more-powerful desk top computers evolve, chemists can expect to see software developers bringing more of the features of the mainframe and minicomputer programs to their PCs.

Product Selection

The programs discussed in this chapter were chosen for their ability to create a 3-D representation of the test structure (*see* page 15). Following are examples of representations produced by some of the programs.

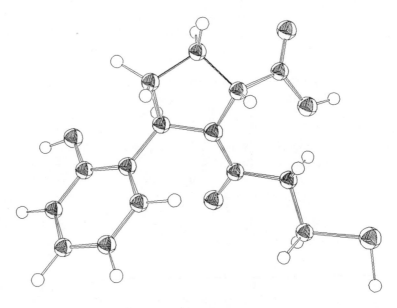

ORTEP representation (PC MODEL)

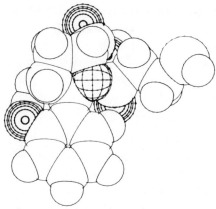

Space-filling model (CAMSEQ/M)

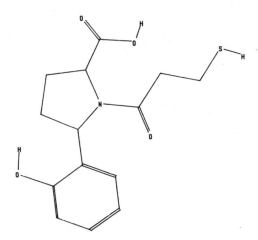

Stick figure (ChemCad)

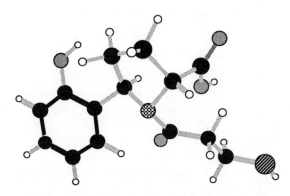

Ball-and-stick representation (Molecular editor VI.I)

Software for Three-Dimensional Molecular Graphics and Modeling

The programs for 3-D molecular graphics and modeling that are discussed in this chapter are the following.
- Alchemy (Tripos Associates)
- CAMSEQ/M (Weintraub Software Design Associates)
- ChemCad (C-Graph Software)
- ChemDraft (C-Graph Software)
- Chem3D (Cambridge Scientific Computing)
- Microchem—Organic Unit (Intersoft)
- Microchem—GAP Unit (Intersoft)
- Microchem—Inorganic Unit (Intersoft)
- Microchem—Macromolecular Unit (Intersoft)
- Modeler—Molecular Design Editor (Queue))
- MOLEC (Queue)
- Molecular Animator (Queue)
- Molecular Editor (Kinko's Academic Courseware Exchange)
- Molecular Graphics (Queue)
- MOLGRAF (Biosoft)
- Molidea (CompuDrug)
- PCDISPLAY (Serena Software)
- PCMODEL (Serena Software)
- XICAMM (Xiris Corporation)

A related program is ChemDraw (*see* Chapter 2).

Product Descriptions

 Alchemy

Alchemy is an interactive, menu-driven molecular-modeling tool for the IBM PC. Much like an electronic Dreiding kit, Alchemy features 3-D molecule construction from atoms, groups, and fragments. The chemist can rotate structures in real-time fashion. Alchemy's energy minimizer uses analytical derivatives to obtain reasonable molecular conformations. Other features are manual docking and fitting of molecules, automatic fitting, geometric measurements, determination of chirality, modification of bond lengths and angles, and stereo image viewing. User-definable atom types and hard-copy output from devices supporting the HPGL standard are also allowed. With 640K of processor memory, the display facility can handle approximately 1000 atoms, and the minimization facility can handle up to 130 atoms.

Hardware

- ☐ PC(s): IBM PCs and compatible systems
- ☐ Graphic board(s): EGA and VGA
- ☐ Mouse(s): Microsoft Mouse and Mouse Systems
- ☐ Printer(s): devices that support the HPGL standard
- ☐ Plotter(s): devices that support the HPGL standard
- ☐ Minimum RAM: 512K (math coprocessor recommended but not required)
- ☐ Dual-floppy-disk system: no

Price and Vendor

- ☐ Industrial price: $750
- ☐ Academic price: $450
- ☐ Vendor: Tripos Associates, Inc., 6548 Clayton Road, St. Louis, MO 63117, (800)323-2960 or (314)647-1099

 ## CAMSEQ/M

CAMSEQ/M is a full-featured, menu-driven, PC-based molecular-modeling system. Its capabilities include a molecular sketchpad that converts user-drawn structures to idealized 3-D molecules; incorporation of templates into structures; several display modes including stick, ball-and-stick, space-filling, and ORTEP displays; full structure inquiry features; dynamic, interactive bond rotations; molecular comparisons and overlays; conformational search; a molecular librarian to assist in organizing structure files; and an interface to other modeling programs, such as MM2. CAMSEQ/M is layered on a device-independent graphics interface to protect the user's hardware investment. It supports most existing displays, input, and hard-copy devices and will support new hardware as it is developed.

Hardware

□ PC(s): IBM PCs and compatible systems
□ Graphic board(s): CGA, EGA, VGA, and Hercules Graphics Card
□ Mouse(s): Microsoft Mouse, Mouse Systems, and Visi-On
□ Graphics tablet(s): Summagraphics
□ Printer(s): Epson FX80 series and compatible systems, Tektronix 4695, and HP LaserJet series and compatible systems
□ Plotter(s): HPGL plotters
□ Minimum RAM: 640K
□ Dual-floppy-disk system: no
□ Version: 2.0

Price and Vendor

□ Industrial price: $795
□ Academic price: $595
□ Vendor: Weintraub Software Design Associates, Inc., P.O. Box 42577, Cincinnati, OH 45242, (513)745-9732

ChemCad

ChemCad is an interactive graphics package that allows the user to design a molecule, optimize its geometry by using MM2 or MNDO calculations, and display and print the results. The program can create large molecules and has a full range of utilities that allows the user to manipulate molecular structures. Geometric parameters such as interatomic distances and angles and dihedral angles can be calculated or user defined. ChemCad also can display an arbitrary number of structures, all of which can be manipulated independently. Preoptimized structure libraries are included with the program to facilitate the rapid construction of molecules. Rotation, zoom, and pan are supported, as well as 3-D ball-and-stick drawings with perspective and hidden-line removal.

Hardware

□ PC(s): IBM PCs and compatible systems
□ Graphic board(s): CGA and EGA
□ Mouse(s): Microsoft Mouse, Mouse Systems, and Logitech
□ Printer(s): Epson FX80 series and compatible systems and HP LaserJet series and compatible systems
□ Plotter(s): HPGL plotters
□ Minimum RAM: 512K

☐ Dual-floppy-disk system: yes
☐ Version: 2.1

Price and Vendor
☐ Industrial price: $250
☐ Academic price: $175
☐ Vendor: C-Graph Software, Inc., P.O. Box 5641, Austin, TX 78763, (512)459-3562

 ChemDraft

ChemDraft is an interactive graphics package that allows the user to create high-quality chemical figures, which may be incorporated into text prepared by a word processor and printed on HP LaserJet-compatible printers. ChemDraft supports 3-D and 2-D molecular structures, a feature that allows the user to construct and manipulate molecules in three dimensions and merge the 3-D structures with 2-D icons (arrows, text, etc.) and structures. The coordinates of a molecule from experimental or computational data can be used, when available, by ChemDraft to display and print the structure as a stick figure or as a 3-D ball-and-stick drawing with perspective and hidden-line removal.

Hardware
☐ PC(s): IBM PCs and compatible systems
☐ Graphic board(s): EGA
☐ Mouse(s): Microsoft Mouse, Mouse Systems, and Logitech
☐ Printer(s): HP LaserJet series and compatible systems
☐ Plotter(s): HPGL plotters
☐ Minimum RAM: 512K
☐ Dual-floppy-disk system: yes
☐ Version: 1.0

Price and Vendor
☐ Industrial and academic price: $295
☐ Vendor: C-Graph Software, Inc., P.O. Box 5641, Austin, TX 78763, (512)459-3562

 Chem3D

Chem3D reveals the 3-D shapes of chemical structures. This molecular-modeling system contains commands to build and view any chemical structure and also to create real-time animation of structures. Building a model with Chem3D can be accomplished in one of three ways: (1) ChemDraw drawings can be understood by Chem3D and fleshed out into 3-D models; (2) Chem3D can read internal, Cartesian, and crystal coordinate files obtained from other sources; or (3) the user can interactively build models with Chem3D's own model builder. Other features include tools to build atom types, a minimization function, and the ability to rotate the model about any axis. Output is available on any printer for which Macintosh printer drivers are available. Pictures of Chem3D models can be integrated into ChemDraw or any package that understands the standard Macintosh PICT format.

Hardware
- ☐ PC(s): Macintosh
- ☐ Graphic board(s): standard Macintosh video cards
- ☐ Mouse(s): standard Macintosh mouse
- ☐ Printer(s): ImageWriter and PostScript printers
- ☐ Minimum RAM: 512K
- ☐ Dual-floppy-disk system: yes
- ☐ Version: 1.0

Price and Vendor
- ☐ Industrial price: $595
- ☐ Academic price: $396
- ☐ Vendor: Cambridge Scientific Computing, Inc., P.O. Box 2123, Cambridge, MA 02238, (617)491-6862

 ## Microchem—Organic Unit

MicroChem consists of a set of separately sold software units covering 3-D organic, inorganic, and macromolecular modeling and physicochemical-property prediction. The Organic Unit consists of the programs InputMol, BuildMol, AssembleMol, DisplayMol, TwistMol, and FormatMol. Organic molecules can be built automatically by joining 3-D fragments or by converting 2-D drawings to 3-D molecules with the energy minimization and geometric construction features of the program. Alternatively, the molecules can be built manually from fractional or Cartesian coordinates input by the user. Conformational energy searching by fixed-valence-geometry molecular mechanics can also be performed. Molecular-graphics capabilities include stick, ball-and-stick, space-filling, and Newman projection representations. Molecule file format conversions are supported for Chemlab-II, SYBYL, and Molecular Design Limited's Molfiles. User-editable parameter and data files are provided.

Hardware

- □ PC(s): Macintosh Plus and SE
- □ Graphic board(s): standard Macintosh video cards
- □ Mouse(s): standard Macintosh mouse
- □ Printer(s): ImageWriter and PostScript printers
- □ Minimum RAM: 1 MB
- □ Dual-floppy-disk system: yes
- □ Version: 2.5

Price and Vendor

- □ Industrial price: $495
- □ Academic price: $295
- □ Vendor: Intersoft, Inc., 1 Concourse Plaza, 4711 Golf Road, Suite 421, Skokie, IL 60045, (312)699-4143

 ## Microchem—GAP Unit

The GAP (Group Additive Properties) Unit of Microchem includes the programs Drug Design and Polymer Transitions and user-editable data files. Classical approaches to molecular-property prediction are useful in situations where 3-D modeling is inconvenient or inapplicable. Drug Design provides a graphical interface to a database of empirically derived property descriptors such as pi, sigma, and molar refractivity. The database can be automatically searched for substituents with one or more properties approximating those of a template substituent, or pairs of properties can be graphically plotted to help systematically vary the choice of analogs for synthesis (a computerized Craig plot). Polymer Transitions predicts T_g (glass transition temperature) and T_m (equilibrium melting temperature) for homo- and copolymers that the user defines from a menu of monomer units.

Hardware

- □ PC(s): Macintosh Plus and SE
- □ Graphic board(s): standard Macintosh video cards
- □ Mouse(s): standard Macintosh mouse
- □ Printer(s): ImageWriter and PostScript printers
- □ Minimum RAM: 1 MB
- □ Dual-floppy-disk system: yes
- □ Version: 2.5

Price and Vendor

- □ Industrial price: $245
- □ Academic price: $145

☐ Vendor: Intersoft, Inc., 1 Concourse Plaza, 4711 Golf Road, Suite 421, Skokie, IL 60045, (312)699-4143

 Microchem—Inorganic Unit

The Inorganic Unit of Microchem consists of the programs ZeoMaker, FormatMol, and DisplayMol. ZeoMaker constructs zeolites (aluminosilicate networks with pores of molecular dimensions) by joining together 3-D units. A limited number of units is supplied with the program. The user can create fragments by using other MicroChem programs (InputMol and BuildMol) or use the FormatMol program to read zeolite files created by other molecular-modeling systems. The user can add or remove framework oxygens, modify the composition (e.g., Si:Al ratio), and assign partial atomic charges to the framework atoms.

Hardware
☐ PC(s): Macintosh Plus and SE
☐ Graphic board(s): standard Macintosh video cards
☐ Mouse(s): standard Macintosh mouse
☐ Printer(s): ImageWriter and PostScript printers
☐ Minimum RAM: 1 MB
☐ Dual-floppy-disk system: yes
☐ Version: 2.5

Price and Vendor
☐ Industrial price: $495
☐ Academic price: $295
☐ Vendor: Intersoft, Inc., 1 Concourse Plaza, 4711 Golf Road, Suite 421, Skokie, IL 60045, (312)699-4143

 Microchem—Macromolecular Unit

The Macromolecular Unit of Microchem consists of the programs SynMacroMol, BioMacroMol, FormatMol, and DisplayMol. SynMacroMol constructs models of synthetic homo- or copolymers including addition, condensation, ladder, or spiro polymers. Syndio-, iso-, and atactic chains can be generated with head-to-tail, tail-to-tail, or random orientations. BioMacroMol constructs models of polyamino acids, polypeptides, and polysaccharides. The macromolecules are generated by automatically propagating the monomer or residue to a selected degree of polymerization. The program can also compute the geometric parameters of helices (e.g., the rise per monomer, angular advancement, and pitch).

Hardware

☐ PC(s): Macintosh Plus and SE
☐ Graphic board(s): standard Macintosh video cards
☐ Mouse(s): standard Macintosh mouse
☐ Printer(s): ImageWriter and PostScript printers
☐ Minimum RAM: 1 MB
☐ Dual-floppy-disk system: yes
☐ Version: 2.5

Price and Vendor

☐ Industrial price: $495
☐ Academic price: $295
☐ Vendor: Intersoft, Inc., 1 Concourse Plaza, 4711 Golf Road, Suite 421, Skokie, IL 60045, (312)699-4143

 Modeler—Molecular Design Editor

Modeler is a molecular-model construction kit that includes color, rotation, and a number of editing facilities. It can be used to display, manipulate, edit, or print out 3-D views of polyatomic structures (up to 50 atoms on the Apple PC and 100 atoms on the IBM PC) by typing in the line structure formula following standard chemical conventions. Modeler allows the user to rotate the model by arbitrary angles about any of the Cartesian axes. It displays the model with atoms numbered for easy identification or with color coding on a color monitor. The user can modify bond lengths and angles to any desired value or use the special twist-and-fuse commands to alter molecular conformation and create complex ring structures. The display operates in static, snapshot, slide, and animation modes.

Hardware

☐ PC(s): IBM PCs and compatible systems and Apple II series
☐ Graphic board(s): CGA and standard Apple II video cards
☐ Mouse(s): not applicable
☐ Printer(s): not applicable
☐ Minimum RAM: 256K on the IBM PC and 48K on the Apple II
☐ Dual-floppy-disk system: yes

Price and Vendor

☐ Industrial and academic price: IBM PC, $95; Apple II, $75
☐ Vendor: Queue, Inc., 562 Boston Avenue, Bridgeport, CT 06610, (800)232-2224 or (203)335-0908

MOLEC

With MOLEC, the user can create models of organic and inorganic molecules containing up to 64 atoms by using either Cartesian coordinates or X-ray crystallographic data. The program disk comes with 23 preprogrammed molecules. MOLEC displays molecules as stick drawings, balls and sticks, or space-filling models. The scale of the molecular display can be changed to blow up portions of the structure. Interatomic angles and distances are measured and displayed on the screen. Molecules can be rotated about the x, y, or z axis by any increment, and an automatic rotation mode simulates real-time rotation. The program includes a screen dump routine to obtain hard-copy pictures of the displayed molecules.

Hardware
□ PC(s): IBM PCs and compatible systems and Apple II series
□ Graphic board(s): CGA and Apple II video card
□ Mouse(s): not applicable
□ Printer(s): not applicable
□ Minimum RAM: 48K
□ Dual-floppy-disk system: yes
□ Version: 1.3

Price and Vendor
□ Industrial and academic price: $75
□ Vendor: Queue, Inc., 562 Boston Avenue, Bridgeport, CT 06610, (800)232-2224 or (203)335-0908

 Molecular Animator

The Molecular Animator uses a combination of hidden-surface techniques and high-speed animation to display and manipulate molecules. It can rotate the molecule on its x, y, or z axis in either direction at any of ten different speeds; rock the structure or stop its rotation in any position; enlarge or reduce its size; or display and rotate the full model—atoms only or a stick model. The program comes with more than 20 molecules ranging from simple organic structures to several complex inorganic compounds. The user can create structures with up to 128 atoms by using either Cartesian or internal coordinates.

Hardware
□ PC(s): IBM PCs and compatible systems and Apple II series
□ Graphic board(s): CGA

☐ Mouse(s): not applicable
☐ Printer(s): not applicable
☐ Minimum RAM: IBM, 256K; Apple, 48K
☐ Dual-floppy-disk system: yes
☐ Version: 1.0

Price and Vendor

☐ Industrial and price: $95
☐ Vendor: Queue, Inc., 562 Boston Avenue, Bridgeport, CT 06610, (800)232-2224 or (203)335-0908

 ## Molecular Editor VI.I

Molecular Editor is a software-based construction kit for building molecules and crystals. The user can build or edit structures of over 100 atoms from any element in the periodic table and use 3-D versions of cut, copy, and paste to build molecules from other structures and functional groups. Structures can be drawn with ball-and-stick, atoms-only, or wire frame models. Molecules can be rescaled, and atomic radii can be independently rescaled while leaving coordinates unchanged. Rotation of a molecule or a portion of a molecule is possible through any angle about any of the three orthogonal axes, either by single steps or continuously. The program can perform group theory operations; measure atomic distances, bond angles, and torsional angles; and change the perspective of the structure. Up to 20 files at one time may be opened and run through rapidly for animation. The program can read and write files in SYLK format as ASCII text, print the structure, and use the extensive help menu and 100+ sample files.

Hardware

☐ PC(s): Macintosh
☐ Graphic board(s): standard Macintosh video cards
☐ Mouse(s): standard Macintosh mouse
☐ Printer(s): ImageWriter and PostScript printers
☐ Minimum RAM: 512K
☐ Dual-floppy-disk system: yes
☐ Version: 1.1

Price and Vendor

☐ Industrial and academic price: $30.35
☐ Vendor: Kinko's Academic Courseware Exchange, 4141 State Street, Santa Barbara, CA 93110-1891, (800)235-6916 or (805)967-0192

Molecular Graphics

With this molecular-display software, the user can create and display with high resolution simple or complex molecules including polymers, proteins, enzymes, or any desired sequence of DNA or RNA. Each of two workspaces can hold structures containing up to 600 atoms. Polymers containing up to 400 residues may be loaded with a special reduce qualifier that displays each residue as a single ball. One or two molecules may be displayed on the same screen, rotated independently, and docked. The program can display wire frame, ball-and-stick, and ball-and-vector (cylindrical bonds) models. Each shaded ball has a unique color and spherical radius that can be altered by the user. The rotation options include global and local rotations and the ability to rotate subgroups or individual bonds. The program also includes online help and search facilities.

Hardware
☐ PC(s): IBM PCs and compatible systems
☐ Graphic board(s): EGA
☐ Mouse(s): not applicable
☐ Printer(s): not applicable
☐ Minimum RAM: 512K (requires a math coprocessor)
☐ Dual-floppy-disk system: yes

Price and Vendor
☐ Industrial price: $500
☐ Academic price: $250
☐ Vendor: Queue, Inc., 562 Boston Avenue, Bridgeport, CT 06610, (800)232-2224 or (203)335-0908

 MOLGRAF

MOLGRAF is a package designed to display, examine, and manipulate 3-D maps of chemical compounds. Its basic features include the entry of new compounds by using crystallographic coordinates and saving data to the disk; the ability to check, add, delete, and correct data; and the calculation of bond lengths, bond angles, and torsion angles. The package comes with an Atlas of data for more than 200 structures, including many common pharmaceuticals. By using MOLGRAF, one can display, rotate, expand, and move structures with high-resolution graphics on the CRT. The images may be printed by using a screen dump program or a plotter. Images of different compounds may be superimposed. Other features include a "lecture" routine to prepare in advance a sequence of images to be viewed and a do-it-yourself facility to build a compound by specifying only bond lengths, bond angles, and torsion angles.

Hardware
☐ PC(s): IBM PCs and compatible systems and Apple II series
☐ Graphic board(s): CGA, EGA, HGA, and standard Apple video card
☐ Mouse(s): not applicable
☐ Printer(s): Epson FX80 series and compatible systems (from screen dump)
☐ Plotter(s): Epson HI-80 plotter
☐ Minimum RAM: IBM PC, 256K; Apple II, 64K
☐ Dual-floppy-disk system: yes

Price and Vendor
☐ Industrial and academic price: $199
☐ Vendor: Biosoft, 68 Hills Road, Cambridge CB2 1LA, England, (44)223 68622, or Biosoft, P.O. Box 580, Milltown, NJ 08850, (201)613-9013

 Molidea

Molidea is an integrated molecular-modeling package that includes model building, molecular mechanics, and quantum chemistry calculations. The molecule-building module automatically calculates Cartesian coordinates, displays a wire frame or space-filled structure on a color graphics screen, and allows rotation of the structure around the x, y, or z axis. The building module can accommodate one molecule of up to 300 H, C, N, O, F, S, Cl, Br, and I atoms. The molecular-mechanics module allows calculation of van der Waals interactions and hydrogen bonding between two or more rigid parts of a molecule or between two or more rigid molecules. The molecular-mechanics module incorporates Lennard–Jones type and optional coulombic potential functions. The quantum chemistry module can perform semiempirical calculations. Molidea can import structures from files in the Molecular Design Limited Molfile format such as MACCS or ChemBase.

Hardware
☐ PC(s): IBM PCs or compatible systems (MS-DOS 3.0 or higher)
☐ Graphic board(s): CGA and EGA
☐ Mouse(s): not applicable; keyboard input
☐ Printer(s): IBM or compatible graphics printer
☐ Minimum RAM: 640K

☐ Dual-floppy-disk system: not recommended
☐ Version: 2.0

Price and Vendor
☐ Industrial price: $1800
☐ Academic price: $900
☐ Vendor: CompuDrug USA, Inc., P.O. Box 202078, Austin, TX 78720, (512)331-0880

 ## PCDISPLAY

PCDISPLAY is a companion program to PCMODEL that provides ORTEP and PLUTO pictures of structures generated by other calculational methods. Input to the program is mouse driven, and the standard MMX, MM2, MNDO, and MOPAC files can be read. A stick structure can be rotated about the x, y, or z axis to obtain the optimum orientation before the ORTEP or PLUTO picture is generated. Hard copy is available from a screen dump or a plot on a Hewlett Packard xy plotter. Structures can also be saved as HPGL files, imported into other drawing programs, and plotted or printed on the HP LaserJet by using a laser plotter program.

Hardware
☐ PC(s): IBM PCs and compatible systems and Macintosh II
☐ Graphic board(s): CGA, EGA, Hercules Graphics Card, and standard Macintosh video cards
☐ Mouse(s): Microsoft Mouse, Mouse Systems, Logitech, and standard Macintosh mouse
☐ Printer(s): HP LaserJet series and compatible systems and Epson FX80 series and compatible systems
☐ Plotter(s): HPGL plotters
☐ Minimum RAM: 400K (requires 8087/80287 numeric coprocessor)
☐ Dual-floppy-disk system: yes

Price and Vendor
☐ Industrial price: $495 (IBM PC version)
☐ Academic price: $200 (IBM PC version)
☐ Vendor: Serena Software, P.O. Box 3076, Bloomington, IN 47402-3076, (812)333-0823

 ## PCMODEL

PCMODEL is a PC molecular-modeling program that features a mouse-driven user interface for structure input, energy minimization, and geometry optimization with the MMX force field for structures of up to

100 atoms and full display options including surfaces and structural information. Current versions support 56 different atom types including several types of carbon, oxygen, nitrogen, sulfur, phosphorus, halogens, transition metals, and transition-state atoms. PCMODEL can read and write MMX, MM2, MOPAC, and MNDO files. This program provides its modeling features in a molecular-mechanics framework and provides a front end to programs that can do more-sophisticated calculations.

Hardware
- ☐ PC(s): IBM PCs and compatible systems and Macintosh II
- ☐ Graphic board(s): CGA, EGA, and Hercules Graphics Cards
- ☐ Mouse(s): Microsoft Mouse, Mouse Systems, and Logitech
- ☐ Printer(s): HP LaserJet series and compatible systems and Epson FX80 series and compatible systems
- ☐ Minimum RAM: 640K (requires 8087/80287 numeric coprocessor)
- ☐ Dual-floppy-disk system: yes

Price and Vendor
- ☐ Industrial price: $495 (IBM PC version)
- ☐ Academic price: $200 (IBM PC version)
- ☐ Vendor: Serena Software, P.O. Box 3076, Bloomington, IN 47402-3076, (812)333-0823

 XICAMM

XICAMM allows structures to be entered, modeled, compared, examined, and stored for later retrieval. The system can calculate bond lengths and angles, dihedral angles, and atomic distances, and it can obtain information about the general strain of atoms and bonds, molecular volumes, principal moments, and connectivity indices. Its modeling system is based on classical molecular mechanics, but it can also handle transition-metal complexes, as well as organic molecules. Graphics input includes a sketch function for drawing rings and chains and retrieval of fragments from disk storage to display structures as stick or space-filled diagrams. Other display modes include stereo view, orthogonal view, and two-structure comparison. The compare function allows any two structures to be directly compared for geometric similarities. The program can handle up to 100 atoms and can generate MM2- and MOPAC-compatible data files.

Hardware
- ☐ PC(s): IBM PCs and compatible systems and Zenith Z248
- ☐ Graphic board(s): CGA, EGA, Hercules Graphics Cards, and any board with DVI drivers
- ☐ Mouse(s): not applicable

☐ Printer(s): Epson FX80 series and compatible systems
☐ Plotter(s): HPGL plotters
☐ Minimum RAM: 640K (requires 8087/80287 math coprocessor)
☐ Dual-floppy-disk system: yes
☐ Version: 2.1

Price and Vendor
☐ Industrial price: $1050 includes graphics drivers
☐ Academic price: $652 includes graphics drivers
☐ Vendor: Xiris Corporation, P.O. Box 787, New Monmouth, New Jersey, (201)671-0517

CHAPTER SIX

Special-Application Software for Chemical Structures

Daniel E. Meyer

S pecial-application software includes programs that allow the uploading and downloading of structural data or the manipulation of structure information for a specific task or function. These programs are usually linked to another product (e.g., online databases and reference books) and provide a user-friendly front end or convenient access to a related product. These programs may also be additional modules for programs covered in the previous four chapters. Special-application programs have been included in this book because of their chemical-graphics capabilities. But because these programs were developed as tools for a specific related product rather than as broad-based utility programs, they are collectively grouped in this final chapter.

Program Features

Of the programs discussed in this chapter, three are directly related to a single online system (CASKit [STN], SuperStructure [CIS], and STN Express [STN]), and one is directly linked to a single group of databases (TOPFRAG [Derwent's patent files]). A fifth program serves as an index aid to the collective volumes of a reference product (SANDRA [Beilstein]). CASKit also converts structure graphics files downloaded from STN into formats usable by other PCs and mainframe computers. All of these programs share the same concept of structure input and use these graphic

1538–3/88/0073/$06.00/0 © 1988 American Chemical Society

data to more conveniently search an online system, database, or data bank.

Two programs (ARGOS and REWARD) convert connection table files into structure diagrams, and a program related to REWARD, DARING, converts Wiswesser line notations (WLN) into connection tables. Another program that uses line notation input is TopKat, which converts SMILES notation into structure displays and then analyzes the compound for a variety of toxicological parameters such as mutagenicity, carcinogenicity, and oral dose that is lethal to 50% of rats (LD_{50}). TopKat also provides data on the probability of a specific toxicity.

An additional module of the Fraser Williams PC software series discussed in this chapter is MARKOUT, which is used to tabulate data collections (both structures and text) and organize the results on the basis of the specific values of a generic structure.

All of the programs discussed in this chapter are written for the IBM PC and compatible systems.

Product Selection

The programs discussed in this chapter were chosen because they use chemical-structure displays. This discussion is made to enable researchers to become aware of the specific designs of the programs and how they differ from the other categories of products that use structure displays.

Special-Application Software

The software for special applications discussed in this chapter are the following.
- ARGOS (Springer–Verlag)
- CASKit-1 (DH Limited)
- CASKit-1 and -2 (DH Limited)
- PC-DARING (Fraser Williams)
- PC-MARKOUT (Fraser Williams)
- PC-REWARD (Fraser Williams)
- SANDRA (Springer–Verlag)
- STN Express (STN International)
- SuperStructure (Fein–Marquart Associates)
- TOPFRAG (Derwent)
- TopKat (Health Designs)

The following are related software discussed elsewhere in this book.

- PsiBase/PsiGen (*see* Chapters 2 and 4)
- TopDog and PC-SABRE (*see* Chapter 4)

Additional special-application software that are available but not included in this survey are MOLKICK and DARC CHEMLINK. MOLKICK is a front-end software for accessing the Beilstein online database and is scheduled for release in August 1988. It is available from Springer–Verlag, 175 Fifth Avenue, New York, NY 10010, (212)460-1500.

DARC CHEMLINK is a front-end software for accessing structure databases on the DARC system and is scheduled for release in 1988. It is available from Questel, Inc., 5201 Leesburg Pike, Falls Church, VA 22041, (703)845-1133, or from Telesystems/Questel, 83–85 Boulevard Vincent Auriol, 75013 Paris, France, (33)1582-6464.

Product Descriptions

 ARGOS

This program automatically converts connection table files that are in an ASCII file format into molecular structural formulas. The information is read sequentially, and the structures are displayed in rectangular fields. The program allows the user to vary the size of the structures; to display more than one structure at a time; and to display stereochemistry, charges, and radicals. ARGOS can be used as a stand-alone program or linked to an organic-synthesis program.

Hardware
☐ PC(s): IBM PCs and compatible systems
☐ Graphic board(s): CGA, EGA, and Hercules Graphics Card
☐ Mouse(s): not applicable
☐ Printer(s): Epson FX80 series and compatible systems and HP LaserJet series and compatible systems
☐ Plotter(s): HPGL plotters
☐ Minimum RAM: 300K
☐ Dual-floppy-disk system: yes
☐ Version: 1.0

Price and Vendor
☐ Industrial and academic price: $2400
☐ Vendor: Springer–Verlag, 175 Fifth Avenue. New York, NY 10010, (212)460-1500

CASKit-1

CASKit-1 is a program for the IBM PC that converts graphic files captured from CAS ONLINE into a clean ASCII file and graphic MDL (Molecular Design Limited) metafiles suitable for use with ChemText. The structures can be scaled and overlaid with numbers, arrows, text, ellipses, arcs, and boxes by using ChemText, and via METAFORM, the size of text characters can also be scaled. The clean ASCII file can be handled by any text processor and output to a laser printer.

Hardware
☐ PC(s): IBM PCs and compatible systems
☐ Graphic board(s): CGA and EGA
☐ Mouse(s): not applicable
☐ Printer(s): Epson FX80 series and compatible systems and PostScript printers
☐ Minimum RAM: 256K
☐ Dual-floppy-disk system: yes
☐ Version: 1.0

Price and Vendor
☐ Industrial and academic price: $75
☐ Vendor: DH Limited, 100 Segre Place, Santa Cruz, CA 95060, (408)427-0321

 CASKit-1 and -2

CASKit-2 completes the path from CAS ONLINE to a PC structure database. CASKit-2 requires CASKit-1 for preprocessing. The results of CAS ONLINE structure searches can be directly stored in personal structure database programs such as ChemBase or MACCS-II. In these structure databases, the results of a literature structure search can be retrieved by exact or substructure match or by search on any other data element. Structures can be moved with their properties through ADAPT, or 3-D models of the structures can be built with PRXBLD followed by Chemlab or MMP2 of Molecular Design Limited.

Hardware
☐ PC(s): IBM PCs and compatible systems
☐ Graphic board(s): CGA and EGA
☐ Mouse(s): not applicable
☐ Printer(s): Epson FX80 series and compatible systems
☐ Minimum RAM: 256K

☐ Dual-floppy-disk system: yes
☐ Version: 1.0

Price and Vendor
☐ Industrial and academic price: $150 (includes both programs)
☐ Vendor: DH Limited, 100 Segre Place, Santa Cruz, CA 95060, (408)427-0321

 PC-DARING

PC-DARING provides personal computer users with the ability to convert Wiswesser line notation (WLN) into connection tables. The program performs the same function as DARING on the mainframe. The program can be integrated with REWARD to provide chemical-structure diagrams and used in conjunction with PC-SABRE to provide structure registration, storage, search, and display facilities.

Hardware
☐ PC(s): IBM PCs and compatible systems
☐ Graphic board(s): not applicable
☐ Mouse(s): not applicable
☐ Printer(s): not applicable
☐ Minimum RAM: 512K
☐ Dual-floppy-disk system: yes
☐ Version: 1.0

Price and Vendor
☐ Industrial and academic price: £575
☐ Vendor: Fraser Williams (Scientific Systems) Ltd., London House, London Road South, Poynton, Cheshire SK12 1YP, England, (44)625-871126

 PC-MARKOUT

PC-MARKOUT is a program that takes an input list of specific chemical structures and data and tabulates the structures and data in terms of a generic structure, which can be either input by the user or derived by the program. Facilities are provided for the user to sort and edit the list of structures.

Hardware
☐ PC(s): IBM PCs and compatible systems
☐ Graphic board(s): CGA, EGA, and Hercules Graphics Card

☐ Mouse(s): not applicable
☐ Printer(s): Epson FX80 series and compatible systems
☐ Minimum RAM: 512K
☐ Dual-floppy-disk system: yes
☐ Version: 1.0

Price and Vendor
☐ Industrial and academic price: £4000
☐ Vendor: Fraser Williams (Scientific Systems) Ltd., London House, London Road South, Poynton, Cheshire SK12 1YP, England, (44)625-871126

 PC-REWARD

PC-REWARD generates two-dimensional structure diagram coordinates from connection tables, such as those produced by PC-DARING. The program contains optional-display and diagram-quality-assessment modules. Diagrams can be transferred to the PC-SABRE search package.

Hardware
☐ PC(s): IBM PCs and compatible systems
☐ Graphic board(s): CGA, EGA, and Hercules Graphics Card
☐ Mouse(s): not applicable
☐ Printer(s): Epson FX80 series and compatible systems
☐ Minimum RAM: 512K
☐ Dual-floppy-disk system: yes
☐ Version: 1.0

Price and Vendor
☐ Industrial and academic price: £550
☐ Vendor: Fraser Williams (Scientific Systems) Ltd., London House, London Road South, Poynton, Cheshire SK12 1YP, England, (44)625-871126

 SANDRA

SANDRA, an acronym for Structure and Reference Analyzer, is a software that analyzes input organic structure diagrams and leads the user to the page range in Beilstein where information on the compound should appear. The program has a menu interface and requires a mouse for graphical structure input. The program analyzes the structure in terms of the Beilstein system and, once the analysis is complete, reports the following information: the series, volume, and subvolume numbers; H-page or Hauptwerk page numbers; system number or range; degree of unsaturation of the registered compound; and carbon number of the registered compound.

Hardware
☐ PC(s): IBM PCs and compatible systems
☐ Graphic board(s): CGA, EGA, and Hercules Graphics Card
☐ Mouse(s): Microsoft Mouse and Mouse Systems
☐ Printer(s): Epson FX80 series and compatible systems
☐ Minimum RAM: 256K
☐ Dual-floppy-disk system: yes
☐ Version: 2.0

Price and Vendor
☐ Industrial price: $500
☐ Academic price: $250
☐ Vendor: Springer–Verlag, 175 Fifth Avenue, New York, NY 10010, (212)460-1500

 ## STN Express

STN Express is front-end software that provides access to STN International structure databases. Specific features of the program include guided search, offline chemical-structure query formulation, and offline search and strategy formulation. The program has special search-and-display features, data capture, and automatic log on. The guided-search feature allows the user to input a search query through a series of menus and then have it automatically processed in the STN database. Off-line chemical-structure building is menu driven and requires the use of a mouse. The structure-input software was derived from the graphic input commands of PsiGen (*see* Chapter 2). In addition to the ability to create search strategies off line, the program provides predefined search strategies for general subjects, such as toxicology, that take advantage of individual databases provided by STN. Graphics are fully integrated with text and can be captured in transcript files and printed.

Hardware
☐ PC(s): IBM PCs and compatible systems
☐ Graphic board(s): CGA, EGA, VGA, and Hercules Graphics Card
☐ Mouse(s): MicroSoft Mouse and Mouse Systems; cursor keys optional
☐ Printer(s): Epson FX8 series and compatible systems, PostScript printers, and HP LaserJet and compatible systems
☐ Minimum RAM: 640K

☐ Dual-floppy-disk system: no
☐ Version: 2.00

Price and Vendor
☐ Industrial price: $595
☐ Academic price: $476
☐ Vendor: STN International—Columbus, P.O. Box 02228, Columbus, OH 43202, (800)848-6538 or (614)447-3600, or STN International—Karlsruhe, Postfache 2465, Karlsruhe D-7500, Federal Republic of Germany, (49)7247-824600

 SuperStructure

SuperStructure permits users to draw chemical structures on a PC with a mouse and then use the structure as an input query to the Structure and Nomenclature Search System (SANSS, a component of the Chemical Information System) to search for identical or similar structures. Super-Structure can also be used by itself to build a private registry system or in conjunction with word processors for technical-report generation.

Hardware
☐ PC(s): IBM PCs and compatible systems
☐ Graphic board(s): CGA, EGA, and Hercules Graphics Card
☐ Mouse(s): Microsoft Mouse and Mouse Systems
☐ Printer(s): Epson FX80 series and compatible systems
☐ Minimum RAM: 512K
☐ Dual-floppy-disk system: yes

Price and Vendor
☐ Industrial and academic price: $99
☐ Vendor: Fein–Marquart Associates, 7215 York Road, Baltimore, MD 21212, (301)821-5980

 TOPFRAG

TOPFRAG converts chemical structures drawn on the PC to the the correct Derwent chemical fragmentation codes and compiles them into fully time-ranged search strategies for direct input to Derwent databases on Orbit, Dialog, or Questel. The automatic conversions provided by the program include tautomer perception and allocation of ring index numbers. The program can also select, through a menu-driven interface, nonstructural concepts such as activities or uses and edit input strategies.

Hardware

☐ PC(s): IBM PCs and compatible systems
☐ Graphic board(s): CGA, EGA, and Hercules Graphics Card
☐ Mouse(s): Microsoft Mouse and Mouse Systems
☐ Printer(s): Epson FX80 series and compatible systems
☐ Minimum RAM: 512K
☐ Dual-floppy-disk system: no
☐ Version: 1.04

Price and Vendor

☐ Industrial and academic price: $675
☐ Vendor: Derwent Publications, Ltd., Rochdale House, 128 Theobalds Road, London WCIX 8RP, England, (44)124-25823, or Derwent Inc., 6845 Elm Street, Suite 500, McLean, VA 22101, (703)790-0400

 TopKat

TopKat is a menu-driven software package for the prediction of toxicity endpoints from the analysis of input chemical structures. TopKat is fully menu driven. TopKat accepts SMILES notation, which is converted into a structure display. The structure can be analyzed against several predictive models, including rat oral LD_{50}, mutagenicity (Ames test), carcinogenicity (NCI/NTP), teratogenicity, rabbit skin irritation, and rabbit eye irritation. The software is intended to be used by toxicologists, pharmacologists, synthetic and medicinal chemists, regulators, and industrial hygienists.

Hardware

☐ PC(s): IBM PCs and compatible systems
☐ Graphic board(s): EGA
☐ Mouse(s): not applicable
☐ Printer(s): Epson FX80 series and compatible systems
☐ Minimum RAM: 640K
☐ Dual-floppy-disk system: no
☐ Version: 1.31

Price and Vendor

☐ Industrial price: $15,000 (program and one module); $5000 for each additional module
☐ Academic price: $11,000 (program and one module); $3750 for each additional module
☐ Vendor: Health Designs, Inc., 183 East Main Street, Rochester, NY 14604, (716)546-1464

APPENDIX ONE

Software by Category and Type of Personal Computer

Software for IBM and Compatible Systems

Structure Drawing	Graphics Terminal Emulation	Structure Management	Chemical Modeling	Special Application
ChemText	ChemTalk	ChemBase	Alchemy	ARGOS
ChemWord	EMU-TEK	CHEMDATA	CAMSEQ/M	CASKit-1
ChiWriter	PC-PLOT-IV Plus	ChemFile II	ChemCad	CASKit-1 and -2
2D-CHEMICAL STRUCTURES	Reflection 7 Plus	ChemSmart	ChemDraft	PC-DARING
MPG	SmarTerm 240	PC-SABRE/PICASSO	Modeler	PC-MARKOUT
PICASSO/DEGAS	TGRAF-05	PsiBase/PsiGen	MOLEC	PC-REWARD
PsiGen	TGRAF-07	TREE (HTSS)	Molecular Animator	SANDRA
Spellbinder Scientific	TGRAF-15		Molecular Graphics	STN Express
T³			MOLGRAF	SuperStructure
TechSet			Molidea	TOPFRAG
TechWriter			PCDISPLAY	TopKat
The Egg			PCMODEL	
WIMP			XICAMM	
WordMarc Composer+				

Software for Macintosh

Structure Drawing	Graphics Terminal Emulation	Structure Management	Chemical Modeling
ChemDraw	TextTerm + Graphics	CHEMDATA	Chem3D
ChemIntosh	TGRAF-07		Microchem—GAP
ChemPlate/Hopkins	TGRAF-15		Microchem—Inorganic
Draw Structures/Organic Fonts	VersaTerm/VersaTerm-PRO		Microchem—Macromolecules
2D-CHEMICAL STRUCTURES			Microchem—Organic
			Molecular Editor
			PCDISPLAY
			PCMODEL

NOTE: There is no special-application software listed for the Macintosh system.

Software for the Apple II Series

- *Structure drawing:* GIOS
- *Graphics terminal emulation:* none listed
- *Structure management:* none listed
- *Chemical modeling:* Modeler, MOLEC, Molecular Animator, and MOLGRAF
- *Special application:* none listed

Software for the UNIX Operating System

- *Structure drawing:* CHEM
- *Graphics terminal emulation:* TGRAF-07
- *Structure management:* none listed
- *Chemical modeling:* none listed
- *Special application:* none listed

APPENDIX TWO

Industrial Prices of Software for Chemical Structures

Table I. Structure-Drawing Software

Price Range (U.S. $)	IBM	Macintosh	Apple II Series	UNIX
≤100		ChemPlate/Hopkins DrawStructures/Organic Fonts		CHEM
101–200	ChiWriter			
201–300	ChemWord MPG TechSet WIMP	ChemIntosh		
301–400	TechWriter			
401–500	PsiGen WordMARC Composer+			
501–600	T³	ChemDraw		
601–700	The Egg			
701–800	Spellbinder Scientific		GIOS[a]	
801–900				
901–1000				
>1000	2-D CHEMICAL STRUCTURES[a] ChemText PICASSO/DEGAS[a]	2-D CHEMICAL STRUCTURES[a]		

[a]Price in foreign currency was converted to U.S. dollars.

Table II. Graphics Terminal Emulation Software

Price Range (U.S. $)	IBM	Macintosh	UNIX
≤100	EMU-TEK-1	VersaTerm	
101–200	PC-PLOT-IV Plus	TextTerm + Graphics	
201–300	EMU-TEK-2	VersaTerm-PRO	
301–400	SmarTerm 240 TGRAF-05		
401–500	EMU-TEK-5 Reflection 7 Plus		
501–600			
601–700	EMU-TEK-7		
701–800			
801–900			
901–1000	ChemTalk TGRAF-07	TGRAF-07	TGRAF-07
>1000	TGRAF-15	TGRAF-15	

NOTE: There is no graphics terminal emulation software listed for the Apple II series.

Table III. Structure Management Software

Price Range (U.S. $)	IBM	Macintosh
≤100		
101–200	ChemSmart	
201–300	ChemFile II	
301–400		
401–500	TREE (HTSS)	
501–600		
601–700		
701–800		
801–900	PsiBase/PsiGen	
901–1000		
>1000	CHEMDATA[a] ChemBase PC-SABRE/PICASSO[a]	CHEMDATA[a]

NOTE: There is no structure management software listed for the Apple II series and the UNIX system.
[a] Price in foreign currency was converted to U.S. dollars.

Table IV. Chemical-Modeling Software

Price Range (U.S. $)	IBM	Macintosh	Apple II Series
≤100	Modeler MOLEC Molecular Animator	Molecular Editor	Modeler MOLEC Molecular Animator
101–200	MOLGRAF		MOLGRAF
201–300	ChemCad ChemDraft	Microchem—GAP	
301–400			
401–500	Molecular Graphics PCDISPLAY PCMODEL	Microchem—Inorganic Microchem—Macromolecules Microchem—Organic PCDISPLAY PCMODEL	
501–600		Chem3D	
601–700			
701–800	Alchemy CAMSEQ/M		
801–900			
901–1000			
>1000	XICAMM Molidea		

NOTE: There is no chemical-modeling software listed for the UNIX system.

Table V. Special-Application Software

Price Range (U.S. $)	IBM
≤100	CASKit-1 SuperStructure
101–200	CASKit-1 and -2
201–300	
301–400	
401–500	SANDRA
501–600	STN Express
601–700	TOPFRAG
701–800	
801–900	
901–1000	PC-DARING[a] PC-REWARD[a]
>1000	ARGOS PC-MARKOUT[a] TopKat

NOTE: There is no special-application software listed for the Macintosh, Apple II, and UNIX systems.
[a]Price in foreign currency was converted to U.S. dollars.

Academic Prices of Software for Chemical Structures

Table I. Structure-Drawing Software

Price Range (U.S. $)	IBM	Macintosh	Apple II Series	UNIX
≤100		Chemplate/Hopkins DrawStructures/Organic Fonts		CHEM
101–200	ChiWriter			
201–300	ChemWord MPG PsiGen TechSet TechWriter WIMP	ChemIntosh		
301–400	ChemText	ChemDraw		
401–500	T³ WordMARC Composer+			
501–600	The Egg			
601–700				
701–800	Spellbinder Scientific		GIOS[a]	
801–900				
901–1000				
>1000	2D-CHEMICAL STRUCTURES[a] PICASSO/DEGAS[a]	2D-CHEMICAL STRUCTURES[a]		

[a]Price in foreign currency was converted to U.S. dollars.

Table II. Graphics Terminal Emulation Software

Price Range (U.S. $)	IBM	Macintosh	UNIX
≤100	EMU-TEK-1	VersaTerm	
101–200	PC-PLOT-IV Plus	TextTerm + Graphics	
201–300	EMU-TEK-2	VersaTerm-PRO	
301–400	SmarTerm 240 TGRAF-05		
401–500	ChemTalk EMU-TEK-5 Plus Reflection 7 Plus		
501–600			
601–700	EMU-TEK-7 Plus		
701–800			
801–900			
901–1000	TGRAF-07	TGRAF-07	TGRAF-07
>1000	TGRAF-15	TGRAF-15	

NOTE: There is no graphics terminal emulation software listed for the Apple II series.

Table III. Structure Management Software

Price Range (U.S. $)	IBM	Macintosh
≤100		
101–200	ChemSmart	
201–300	ChemFile II	
301–400	TREE (HTSS)	
401–500	PsiBase/PsiGen	
501–600		
601–700		
701–800		
801–900		
901–1000	ChemBase	
>1000	CHEMDATA[a] PC-SABRE/PICASSO[a]	CHEMDATA[a]

NOTE: There is no structure management software listed for the Apple II series and the UNIX system.
[a]Price in foreign currency was converted to U.S. dollars.

Table IV. Chemical-Modeling Software

Price Range (U.S. $)	IBM	Macintosh	Apple II Series
≤100	Modeler MOLEC Molecular Animator	Molecular Editor	Modeler MOLEC Molecular Animator
101–200	ChemCad MOLGRAF PCDISPLAY PCMODEL	Microchem—Gap PCDISPLAY PCMODEL	MOLGRAF
201–300	ChemDraft Molecular Graphics	Microchem—Inorganic Microchem—Organic Microchem—Macromolecules	
301–400	Alchemy	Chem3D	
401–500			
501–600	CAMSEQ/M		
601–700	XICAMM		
701–800			
801–900	Molidea		
901–1000			
>1000			

NOTE: There is no chemical-modeling software listed for the UNIX system.

Table V. Special-Application Software

Price Range (U.S. $)	IBM
≤100	CASKit-1 SuperStructure
101–200	CASKit-1 and -2
201–300	SANDRA
301–400	
401–500	STN Express
501–600	
601–700	TOPFRAG
701–800	
801–900	
901–1000	PC-DARING[a] PC-REWARD[a]
>1000	ARGOS PC-MARKOUT[a] TopKat

NOTE: There is no special-application software listed for the Macintosh, Apple II, and UNIX systems.
[a]Price in foreign currency was converted to U.S. dollars.

Glossary

Ames test in vitro bioassay for mutagenicity using microorganisms

ASCII American Standard Code for Information Interchange; a seven-bit code used by computers to represent the common keyboard characters (letters of the alphabet, numbers, and punctuation marks) and control codes

AWK pattern language in UNIX used as a preprocessor; the letters stand for the AT&T Bell Laboratory scientists who developed it: Alfred Aho, Peter Weinberger, and Brian Kernigham

bit-mapped images graphics that are stored as discrete dots and that can be edited by manipulating discrete pixel points on the computer screen; also called raster images

BRS Bibliographic Retrieval Service; provided by BRS Information Technologies, an online host of scientific, medical, and business databases

CAD/CAM computer-aided design/computer-aided manufacturing

CAS ONLINE online system for accessing the *Chemical Abstracts* database on STN International

CD-ROM compact disk read only memory; a 5 1/4-inch optical disk used to store large amounts of text and graphics data

CGA color graphics adapter; low-resolution graphics card for the IBM and compatible computers

97

CIS Chemical Information System; developed by the National Institutes of Health and the Environmental Protection Agency and now supported by Fein–Marquart Associates

clipboard standard Macintosh feature that allows users to transfer graphics and text between different programs

connection table machine-readable format for chemical structures which stores information about each atom in a structure such as atomic symbol; x, y, and z coordinates; connectivity to other atoms; type of bonding (single, double, triple, normalized, etc.); and sequence number. The table may also contain other information such as valency, charge, isotopic mass, and direction of bonding. This information is used to later retrieve and display the structure.

CPU central processing unit; the part of a computer that performs calculations and determines where data is read and stored

CROSSBOW Computerized Retrieval of Organic Structures Based On Wiswesser line notation input; chemical-structure-handling system developed by ICI and marketed by Fraser Williams

CRT cathode ray tube; the part of a monitor that is the screen

DARC description, acquisition, retrieval, and correlation; chemical-structure-handling software developed by Dubois at the University of Paris, France, and marketed by Telesystems Questel

Dialog online host of more than 300 databases marketed by Dialog Information Services, Inc.

DVI device independent; program output that is independent of any one device, for example a screen or a printer

EGA enhanced graphics adapter; graphics card for the IBM and compatible computers

EQN preprocessing language for *troff* used to specify equations

HGA Hercules graphics adapter; graphics card for the IBM and compatible computers

HPGL Hewlett–Packard graphics language; plotter language for Hewlett–Packard and compatible plotters

K kilobyte

Kermit public-domain file-transfer protocol supported by a variety of computers

LAN local area network; hardware and software connection among several computers that allows these computers to share files and programs

MACCS Molecular Access System; software for handling chemical structures on minicomputers and mainframes; developed by Molecular Design Limited

MB megabyte

MM2 Molecular Mechanics II; program used for the calculation of conformations and energies

MNDO modified neglected differential overlap; program that performs calculations, such as geometry and energy, by using molecular orbital theory

Molfile file format for the storage of graphics used by Molecular Design Limited programs

MOPAC general molecular orbital package that performs geometric optimizations and calculations of vibrational frequencies and thermodynamic quantities

MS-DOS Microsoft Disk Operating System; operating system for IBM-compatible computers developed by Microsoft Corporation

object-oriented graphics graphics that are drawn by simple primary objects such as circles, rectangles, lines, and arcs rather than as bit-mapped images

ORTEP Oak Ridge thermal ellipsoid plot

PC-DOS Personal Computer Disk Operating System; operating system specifically for IBM computers

PDIP proprietary file-transfer protocol developed by Persoft

PIC preprocessing language for *troff* used to specify schemes and diagrams

PICT file format used by Macintosh computers to store graphics files produced by QuickDraw or by object-oriented graphics programs that use the QuickDraw routines

poly-gonfil computer routine used to fill outlined geometric objects such as triangles, squares, and circles

RAM random access memory; used by a computer to store programs that are being executed and data temporarily

REACCS Reaction Access System; software for handling reaction information on minicomputers and mainframes; developed by Molecular Design Limited

ROM read only memory; used to store unchanging program instructions, for example those programs that are part of the operating system for the computer

SMILES line notation for representing chemical structures; developed at Pomona College, Claremont, CA

STN Scientific and Technical Network; international online network offered cooperatively by the American Chemical Society (ACS), Fachinformationszentrum Energie, Physik, Mathematik GmbH (FIZ Karlsruhe), and the Japan Association for International Chemical Information (JAICI)

SYLK symbolic link; file format that represents a worksheet, for example a spreadsheet or connection tables, which allows the information to be exchanged between programs

TBL preprocessing language for *troff* used to specify tables

troff text-formatting program for typesetting machines developed for the UNIX operating system by AT&T Bell Laboratories

TTY teletype terminal

UNIX operating system developed by AT&T Bell Laboratories

VGA video graphics array; graphics adapter for the IBM and compatible computers

WLN Wiswesser line notation; linear coding system for representing two-dimensional chemical structures; developed by William Wiswesser

WORM write once read many; optical disk that can be loaded with data once by the user and read many times

WYSIWYG what you see is what you get; the ability of some programs to display on the screen what the program will produce on hard-copy output

Xmodem public-domain file-transfer protocol supported by a variety of computers

Index